T0142072

Numbers and Computers

Ronald T. Kneusel

Numbers and Computers

 Springer

Ronald T. Kneusel
Broomfield, CO, USA

ISBN 978-3-319-35940-3   ISBN 978-3-319-17260-6   (eBook)
DOI 10.1007/978-3-319-17260-6

Springer Cham Heidelberg New York Dordrecht London

Printed on acid-free paper

Springer International Publishing AG Switzerland is part of Springer Science+Business Media (www.
springer.com)

To my parents, Janet and Tom, for fostering my love of science.

Preface

This is a book about numbers and how those numbers are represented in and operated on by computers.

Of course, numbers are fundamental to how computers operate because, in the end, everything a computer works with is a number. It is crucial that people who develop code understand this area because the numerical operations allowed by computers and the limitations of those operations, especially in the area of floating point math, affect virtually everything people try to do with computers. This book aims to help by exploring, in sufficient, but not overwhelming, detail, just what it is that computers do with numbers.

Who Should Read This Book

This book is for anyone who develops software including software engineers, scientists, computer science students, and engineering students and anyone who programs for fun.

If you are a software engineer, you should already be familiar with many of the topics in this book, especially if you have been in the field for any length of time. Still, I urge you to press on, for perhaps you will find a gem or two which are new to you.

If you are old enough, you will remember the days of Fortran and mainframes. If so, like the software engineers above, you are probably also familiar with the basics of how computers represent and operate on numbers, but, also like the software engineers above, you will likely find a gem or two of your own. Scientists in particular should be aware of the limitations and pitfalls associated with using floating point numbers since few things in science are restricted to integers.

Students need this book because it is essential to know what the computer is doing under the hood. After all, if you are going to make a career of computers, why would you not want to know how the machine works?

How to Use This Book

This book consists of two main parts. The first deals with standard representations of integers and floating point numbers, while the second details several other number representations which are nice to know about and handy from time to time. Either part is a good place to start, though it is probably best if the parts themselves are read from start to end. Later, after the book has been read, you can use it as a reference.

There are exercises at the end of each chapter. Most of these are of the straightforward pencil and paper kind, just to test your understanding, while others are small programming projects meant to increase your familiarity with the material. Exercises that are (subjectively) more difficult will be marked with either one or two stars (* or **) to indicate the level of difficulty.

Example code is in C and/or Python version 2.7, though earlier 2.x versions should work just as well. Intimate knowledge of these programming languages is not necessary in order to understand the concepts being discussed. If something is not easy to see in the code, it will be described in the text. Why C? Because C is a low-level language, close to the numbers we will be working with and because C is the grandfather of most common programming languages in current use including Python. In general, code will be offset from text and in a monospace font. For readers not familiar with C and/or Python, there are a plethora of tutorials on the Web and reference books by the bookcase. Two examples, geared toward people less familiar with programming, are *Beginning C* by Ivor Horton and *Python Programming Fundamentals* by Kent Lee. Both of these texts are available from Springer in print or e-book format.

At the end of each chapter are references for the material presented in the chapter. Much can be learned by looking at these references. Almost by instinct we tend to ignore sections like this as we are now programmed to ignore advertisements on web pages. In this former case, resist temptation; in the latter case, keep calm and carry on.

Acknowledgments

This book was not written in a vacuum. Here I want to acknowledge those who helped make it a reality. First, the reviewers, who gave of their time and talent to give me extremely valuable comments and friendly criticism: Robert Kneusel, M.S., Jim Pendleton, Ed Scott, Ph.D., and Michael Galloy, Ph.D. Gentlemen, thank you. Second, thank you to Springer, especially my editor, Courtney Clark, for moving ahead with this book. Lastly, and most importantly, thank you to my wife, Maria, and our children: David, Peter, Paul, Monica, Joseph, and Francis. Without your patience and encouragement, none of this would have been written.

Broomfield, USA   Ronald T. Kneusel
December 2014   AM+DG

Contents

Part I
Standard Representations

Chapter 1
Number Systems

Abstract Computers use number bases other than the traditional base 10. In this chapter we take a look at number bases focusing on those most frequently used in association with computers. We look at how to construct numbers in these bases as well as how to move numbers between different bases.

1.1 Representing Numbers

The ancient Romans used letters to represent their numbers. These are the "Roman numerals" which are often taught to children,

$$
\begin{array}{ll}
\text{I} & 1 \\
\text{II} & 2 \\
\text{III} & 3 \\
\text{IV} & 4 \ (1 \text{ before } 5) \\
\text{V} & 5 \\
\text{X} & 10 \\
\text{L} & 50 \\
\text{C} & 100 \\
\text{D} & 500 \\
\text{M} & 1000 \\
\end{array}
$$

By grouping these numbers we can build larger numbers (integers),

$$\text{MCMXCVIII} = 1998$$

The Romans built their numbers from earlier Egyptian numbers as seen in Fig. 1.1. Of course, we do not use either of these number systems for serious computation today for the simple reason that they are hard to work with.

For example, the Egyptians multiplied by successive doubling and adding. So, to multiply 17×18 to get 306 the Egyptians would make a table with two columns. The first column consists of the powers of two: $2^0, 2^1, 2^2, 2^3, 2^4 = 1, 2, 4, 8, 16$, etc.

R.T. Kneusel, *Numbers and Computers*, DOI 10.1007/978-3-319-17260-6_1

$$| = 1 \qquad \cap = 10$$

$$|| = 2 \qquad \cap\cap = 20$$

$$||| = 3 \qquad \cap\cap\cap = 30$$

$$|||| = 4 \qquad \mathcal{9} = 100$$

$$||||\,\cap\cap\,\mathcal{9}\mathcal{9} = 224$$

Fig. 1.1 Egyptian numbers. The ancient Egyptians wrote numbers by summing units, tens, and hundreds. There were also larger valued symbols which are not shown. Numbers were written by grouping symbols until the sum equaled the desired number. The order in which the symbols were written, largest first or smallest first, did not matter, but common symbols were grouped together. In this example, the smallest units are written first when writing from left to right to represent 224

and the second column is the first number to be multiplied and its doublings. In this case, 17 is the first number in the problem so the second column would be $17, 34, 68, 136, 272$, etc. where the next number in the sequence is two times the previous number. Therefore, the table would look like this (ignore the starred rows for the moment),

$$
\begin{array}{rl}
1 & 17 \\
2 & 34 \quad * \\
4 & 68 \\
8 & 136 \\
16 & 272 \;*
\end{array}
$$

Next, the Egyptians would mark the rows of the table that make the sum of the entries in the first column equal to the second number to multiply, in this case 18. These are marked in the table already and we see that $2 + 16 = 18$. Lastly, the final answer is found by summing the entries in the second column for the marked rows, $34 + 272 = 306$.

This technique works for any pair of integers. If one were to ask the opposite question, what number times 17 gives 306, one would construct the table as before and then mark rows until the sum of the numbers in the second column is 306.

The answer, then, is to sum the corresponding numbers in the first column to get 18. Hence, division was accomplished by the same trick. In the end, since we have algebra at our disposal, we can see that what the Egyptians were really doing was applying the distributive property of multiplication over addition,

$$17(18) = 17(2+16) = 34+272 = 306$$

But, still, this is a lot of effort. Fortunately, there is another way to represent numbers: place notation (or, more formally, place-value notation). In place notation, each digit represents the number of times that power of the base is present in the number with the powers increasing by one for each digit position we move to the left. Negative powers of the base are used to represent fractions less than one. The digit values we typically use range from 0 to 9 with 10 as the base. Here we write 10 to mean $1 \times 10^1 + 0 \times 10^0$ thereby expressing the base of our numbers in the place notation using that base. Naturally, this fact is true regardless of the base, so 10_B is the base B for all $B > 1$. Computers do use base 10 numbers but not very often though we will see a few examples of this later in the book.

In general, if the base of a number system is B then numbers in that base are written with digits which run from 0 to $B-1$. These digits, in turn, count how many instances of the base raised to some integer power are present in the number. Lots of words, but an equation may clarify,

$$abcd_B = a \times B^3 + b \times B^2 + c \times B^1 + d \times B^0$$

for digits a, b, c, and d. Notice how the exponent of the base is counting down. This counting continues after zero to negative exponents which are just fractions,

$$B^{-n} = \frac{1}{B^n}$$

So, we can introduce a "decimal point" to represent fractions of the base. Formally, this is known as the *radix point* and is the character which separates the integer part of a number in a particular base from the fractional part. In this book, we will use the "." (period) character and indicate the base of the number with a subscript after the number,

$$abcd.efg_B = a \times B^3 + b \times B^2 + c \times B^1 + d \times B^0 + e \times B^{-1} + f \times B^{-2} + g \times B^{-3}$$

for additional digits e, f, g and base B. If no explicit base is given, assume the base is 10.

Before moving on to modern number systems let's take a quick look at two ancient number systems that used place notation. The first is the Babylonian number system which used base 60 (*sexagesimal*) but constructed its digits by grouping sets of symbols representing ones and tens. The second is the Mayan number system which used base 20 and constructed digits by combining dots for one and bars for five.

Figure 1.2 illustrates Babylonian numbers and performs some simple additions with them. One interesting thing to note is that the ancient Babylonians did not have a symbol for zero or the radix point. Instead, they used spaces and context from the text of the document to imply where the radix point should be inserted. Figure 1.2b shows examples of decimal numbers written in base 60 and as the Babylonians would have written them. Since it is difficult to work with the actual notation, scholars use a shorthand which combines our decimal notation with base 60. In this notation, a number is written with "," (comma) separating the digits and, if necessary, a ";" (semicolon) to serve as the radix point. For example,

$$1234_{10} = 20 \times 60^1 + 34 \times 60^0 = 20,34$$

or

$$3.14159 = 3 \times 60^0 + 8 \times 60^{-1} + 29 \times 60^{-2} + 44 \times 60^{-3}$$

Fig. 1.2 (**a**) Sexagesimal numbers in ancient Babylon used groups of symbols for one and ten to be the digits of their base 60 numbers as opposed to our use of ten different symbols for our digits. (**b**) Decimal numbers written in sexagesimal. The left column is the decimal number, the middle column is the equivalent base 60 number using decimal digits and "," to separate the digits. The right column shows the Babylonian number which matches directly with the middle column representation. (**c**) The Babylonians were able to write numbers with fractional parts by using a space to separate it from the integer part. Here we show a three place representation of π

The fact that the ancient Babylonians had a place notation is all the more impressive given that they wrote and performed their calculations on clay tablets which were allowed to dry in the sun. This is a good thing as it preserved thousands of

Fig. 1.3 YBC 7289 showing the calculation of the diagonal of a square with a side of length $\frac{1}{2}$. Note that this calculation makes use of a particularly good value for $\sqrt{2}$ (Image copyright Bill Casselman and is used with permission, see http://www.math.ubc.ca/~cass/Euclid/ybc/ybc.html. The original tablet is in the Yale Babylonian Collection)

tablets for us to investigate today. Figure 1.3 is a particularly important mathematical tablet. This tablet shows a square with its diagonals marked. The side of the square is marked as 30 which is interpreted to be $\frac{1}{2}$. There are two numbers written by one of the diagonals. The first is 1, 24, 51, 10 and the one below it is 42, 25, 35. The length of the diagonal of a square is $\sqrt{2}$ times the length of the side. If we multiply 30 by 1, 24, 51, 10 we do get 42, 25, 35. This means that the Babylonians used 1, 24, 51, 10 as an approximation to $\sqrt{2}$. How good was this approximation? Squaring 1, 24, 51, 10 gives 1, 59, 59, 59, 38, 1, 40 which is, in decimal, about 1.99999830, a very good approximation indeed.

The Mayan people developed their own place notation in base 20 using combinations of dots (one) and bars (five) to form digits. The numbers were written vertically from bottom to top of the page. Figure 1.4 shows on the left an addition problem written with this notation. Unlike the Babylonians, the Mayans did use a symbol for zero, shaped like a shell, and this is seen in Fig. 1.4 on the right.

Fig. 1.4 Mayan numbers in base 20 written from bottom to top. Bars represent five, dots represent ones, with groups of bars and dots used to show the numbers 1 through 19. The zero symbol, shown on the right, is a shaped like a shell and holds empty places in the same way zero is used in base 10 so that the number on the right is $1 \times 20^3 + 0 \times 20^2 + 0 \times 20^1 + 14 \times 20^0 = 8014$

It is interesting to see how the Maya, like the Babylonians before them, did not create unique symbols for all possible base 20 digits but instead used combinations likely base symbols. This might imply that the place notation evolved from an early tally system, which is a reasonable thing to imagine. Addition in the Mayan system worked by grouping digits, replacing five dots with a bar and carrying to the next higher digit if the total in the lower digit was twenty or more.

For more background on the fascinating history of numbers and number systems, including Mayan numbers, see [1] and [3]. For a detailed look at Babylonian numbers and mathematics, see [4] and [5]. Lastly, for a look at Egyptian numbers and mathematics see [2].

1.2 The Big Three (and One Old Guy)

In this book we are concerned, primarily, with three number bases: base 2 (binary), base 10 (decimal), and base 16 (hexadecimal). These are the "big three" number bases when it comes to computers. In the past, base 8 (octal) was also used frequently, so we will mention it here. This is the "one old guy". In this section we take a look at how to represent numbers in these bases and how to move a number between these bases. Operations, such as arithmetic, will be covered in subsequent chapters from the point of view of how the computer works with numbers of a particular type.

Decimal Numbers. Relatively little needs to be said about decimal numbers as we have been using them all our lives to the point where working with them is second nature. Of course, the reason for this frequent appearance of ten is that we have ten fingers and ten toes. Decimal numbers, however, are not a good choice for computers. Digital logic circuits have two states, "on" or "off". Since computers are built of these circuits it makes sense to represent the "on" state (logical `True`) as a 1 and the "off" state (logical `False`) as a 0. If your digits are 0 and 1 then the natural number base to pick is 2, or binary numbers.

Binary Numbers. In the end, computers work with nothing more than binary numbers which is really quite interesting to think about: all the activities of the computer rely on numbers represented as ones and zeros. Following the notation above, we can see that a binary number such as 1011_2 is really,

$$1011_2 = 1 \times 2^3 + 0 \times 2^2 + 1 \times 2^1 + 1 \times 2^0 = 11$$

Thinking back to the example of ancient Egyptian multiplication above we can now see that the left-hand column was nothing more than the powers of two, which are the place values in a binary number. Since each power of two appears only once in a given number (integer) it is always possible to find a sum of powers of two that equals any number. The 1 values in a binary integer are nothing more than tick marks indicating which power of two is present. We know that each power of two appears at most once in a given integer because it is possible to represent any integer in binary and the only digits allowed in a binary number are 0 and 1.

Hexadecimal Numbers. If you work with computers long enough, you will eventually encounter hexadecimal numbers. These are base 16 numbers since "hex" refers to six and "dec" refers to ten. One question might immediately present itself: if we need ten separate symbols for the ten decimal digits, we need 16 symbols for hexadecimal digits. Zero through nine can be reused, but what should we use for 10 through 15? The flippant answer is, whatever symbols you want, you could even make up sixteen entirely new symbols, but typically we follow the convention the ancient Greeks used (see [1], chapter 4) and take the first six letters of the Latin alphabet as the missing symbols,

$$
\begin{array}{ll}
A & 10 \\
B & 11 \\
C & 12 \\
D & 13 \\
E & 14 \\
F & 15
\end{array}
$$

In this case, counting runs as $1, 2, 3, 4, 5, 6, 7, 8, 9, A, B, C, D, E, F, 10$. It should be no surprise that when writing hexadecimal numbers one often sees lowercase letters,

a–f, instead of A–F. Additionally, computer scientists, like most English speakers, tend to shorten words whenever possible. So, you will often hear or see hexadecimal numbers referred to as "hex" numbers. No reference to magic is intended unless it is the magic by which computers work at all.

If we break down a hex number,

$$FDED_{16} = 15 \times 16^3 + 13 \times 16^2 + 14 \times 16^1 + 13 \times 16^0 = 65005$$

we see that it is a compact notation compared to binary. This should not be surprising as in place of simply indicating which power of two is in a number, as binary does, we are here allowed up to sixteen possible multipliers on each power of 16.

Octal Numbers. While much less common than in the past, octal (base 8) numbers still show up from time to time. Since these are base 8 numbers we already know that we only need the digits 0–7 to write numbers in this base. In Sect. 1.3 we will see why, if computers are really working in binary, people bother to use other number bases like octal or hexadecimal.

We should expect by now that octal numbers break down in a familiar way,

$$1234_8 = 1 \times 8^3 + 2 \times 8^2 + 3 \times 8^1 + 4 \times 8^0 = 668$$

and we will not be disappointed. In modern programming languages, octal numbers are typically reserved for character codes and will be rarely seen outside of that context.

1.3 Converting Between Number Bases

Representing numbers is a good thing but sometimes we need to convert a number between bases. In this section we look at converting between the bases we introduced above. Table 1.1 will be handy when converting numbers between bases. It shows the first sixteen integers in binary, octal, hexadecimal and decimal.

Hexadecimal and Binary. Hexadecimal numbers are base 16 while binary numbers are base 2. We see that $2^4 = 16$ so the conversion between hexadecimal and binary, either direction, will involve groups of four binary digits. To convert a binary number to hexadecimal, simply separate the digits into groups of four, from right to left, and write each group as a hex digit referring to Table 1.1 if necessary,

$$
\begin{aligned}
1011100100111110_2 &= \quad 101\ 1100\ 1001\ 1110 \\
&= \qquad 5 \quad\ \ C \quad\ \ 9 \quad\ \ E \\
&= 5C9E_{16}
\end{aligned}
$$

Turning a hexadecimal number into binary is simply the reverse process,

$$
\begin{aligned}
FDDA_{16} &= \quad F \quad\ D \quad\ D \quad\ A \\
&= 1111\ 1101\ 1101\ 1010 \\
&= 1111110111011010_2
\end{aligned}
$$

Decimal	Binary	Octal	Hexadecimal
0	0000	0	0
1	0001	1	1
2	0010	2	2
3	0011	3	3
4	0100	4	4
5	0101	5	5
6	0110	6	6
7	0111	7	7
8	1000	10	8
9	1001	11	9
10	1010	12	A
11	1011	13	B
12	1100	14	C
13	1101	15	D
14	1110	16	E
15	1111	17	F

Table 1.1 The first sixteen integers in decimal, binary, octal, and hexadecimal

This conversion process is simple enough, but why does it work? The key is in the fact already noted, that $2^4 = 16$. Recall that in a place notation system each digit position to the left is one power of the base larger just as the hundreds place in decimal is a factor of ten, the base, larger than the tens place. If we look at base 2 and instead of moving one digit to the left, which is two times larger than the original digit value, we move four digits we are now $2 \times 2 \times 2 \times 2$ larger or 16. But, in base 16, each digit position is 16 times larger than the next place to the right. Therefore, we have an equivalence between moving one digit to the left in a hexadecimal number and four digits to the left in a binary number. This is why we can group the binary digits in sets of four to get the equivalent hexadecimal digit.

Octal and Binary. If $2^4 = 16$ implies we group binary digits in sets of four, then $2^3 = 8$ implies we should group binary digits in sets of three to convert between binary and octal, and this is indeed the case,

$$
\begin{aligned}
101110010011110_2 &= \quad 101\ 110\ 010\ 011\ 110 \\
&= \quad\ \ 5\ \ \ 6\ \ \ 2\ \ \ 3\ \ \ 6 \\
&= 56236_8
\end{aligned}
$$

Likewise, we can reverse the process to get,

$$
\begin{aligned}
52431_8 &= \quad 5\ \ \ 2\ \ \ 4\ \ \ 3\ \ \ 1 \\
&= 101\ 010\ 110\ 011\ 001 \\
&= 101010110011001_2
\end{aligned}
$$

Lastly, should the need arise, conversion between hexadecimal and octal, or vice versa, is easily accomplished by first moving to binary and then to the other base from binary.

Decimal and Binary. The simple conversion technique above falls apart when moving between decimal and binary. We already know the reason, $2^x = 10$ does not have an integer solution so we cannot simply group digits. Instead, we handle the conversions in a more cumbersome way.

Decimal to Binary. Perhaps the simplest way to convert a decimal number to binary by hand is to repeatedly divide the number by two and keep track of the quotient and remainder. The quotient gives the next number to divide by two and the remainder, always either 1 or 0, is one binary digit of the result. If the numbers are not too big, dividing by two is easy enough to do without paper and pencil.

For example, to convert 123 to binary,

$$
\begin{aligned}
123 \div 2 &= 61 \text{ r } 1 \\
61 \div 2 &= 30 \text{ r } 1 \\
30 \div 2 &= 15 \text{ r } 0 \\
15 \div 2 &= 7 \text{ r } 1 \\
7 \div 2 &= 3 \text{ r } 1 \\
3 \div 2 &= 1 \text{ r } 1 \\
1 \div 2 &= 0 \text{ r } 1
\end{aligned}
$$

Now, to get the binary representation, read from bottom to top: $123 = 1111011_2$.

The conversion works but it is good to know *why* it works so let's take a closer look. If we write a seven digit binary number, $b_6 b_5 b_4 b_3 b_2 b_1 b_0$, with b_k the k-th digit, in expanded form we get,

$$b_0 \times 2^0 + b_1 \times 2^1 + b_2 \times 2^2 + b_3 \times 2^3 + b_4 \times 2^4 + b_5 \times 2^5 + b_6 \times 2^6$$

which, since $2^0 = 1$ and every other term is a multiple of two, can be written as,

$$b_0 + 2(b_1 \times 2^0 + b_2 \times 2^1 + b_3 \times 2^2 + b_4 \times 2^3 + b_5 \times 2^4 + b_6 \times 2^5)$$

If we divide the above by 2 we get the quotient,

$$b_1 \times 2^0 + b_2 \times 2^1 + b_3 \times 2^2 + b_4 \times 2^3 + b_5 \times 2^4 + b_6 \times 2^5$$

and a remainder of b_0. Therefore, it is clear that performing an integer division of a number by two leads to a new quotient and a remainder of either zero or one since b_0 is a binary digit. Further, this remainder is in fact the right-most digit in the binary representation of the number. If we look at the quotient above it has the same form as the initial binary representation of the number but with every binary digit now shifted to the next lowest power of two. Therefore, if we take this quotient and divide it by two we will repeat the calculation above and get b_1 as the remainder. This is then the next digit in the binary representation of the number. Lastly, if we repeat this over and over until the quotient is simply zero we will generate all the digits in the binary representation of the number in reverse order. Finally, writing the stream of binary digits backwards will put them in the proper sequence and give the desired binary representation.

Another method for converting a decimal number to binary involves making a table of the powers of two and asking repeated questions about sums from that table. This method works because the conversion process is really tallying the powers of two that add up to the decimal number.

For example, to convert 79 to binary make a table like this,

64	32	16	8	4	2	1

with the powers of two written highest to lowest from left to right. Next, ask the question, "Is $64 > 79$?" If the answer is yes, put a 0 under the 64. If the answer is no, put a 1 under the 64 because there is a 64 in 79,

64	32	16	8	4	2	1
1						

Now add a new line to the table and ask the question "Is $64 + 32 > 79$?" Since the answer is yes, we know that there is no 32 in the binary representation of 79 so we put a 0 below the 32,

64	32	16	8	4	2	1
1						
	0					

We continue the process, now asking "Is $64 + 16 > 79$?" In this case, the answer is again yes so we put a 0,

64	32	16	8	4	2	1
1						
	0					
		0				

We continue for each remaining power of two asking whether that power of two, plus all the previous powers of two that have a 1 under them, is greater or less than 79,

64	32	16	8	4	2	1
1						
	0					
		0				
			1			
				1		
					1	
						1

To complete the conversion we collapse the table vertically and write the binary answer, $79 = 1001111_2$.

Binary to Decimal. If a binary number is small the fastest way to convert it to a decimal is to simply remember Table 1.1. Barring that, there is a straightforward iterative way to convert binary to decimal. If we have a k digit binary number, $b_4b_3b_2b_1b_0$, where in this case $k = 5$, we can use the following recurrence relation to find the decimal equivalent,

$$d_0 = 0$$
$$d_i = b_{k-i} + 2d_{i-1}, i = 1\ldots k$$

For example, let's convert 11001_2 to decimal using the recurrence relation. In this case we get,

i	d_i
0	0
1	$1 + 2(0) = 1$
2	$1 + 2(1) = 3$
3	$0 + 2(3) = 6$
4	$0 + 2(6) = 12$
5	$1 + 2(12) = 25$

This technique is easily implemented in a computer program. Why does it work? Consider the five digit binary number we started this section with, $b_4b_3b_2b_1b_0$. If we write the algebraic expression that represents the recurrence relation after five iterations we get,

$$d_5 = b_0 + 2(b_1 + 2(b_2 + 2(b_3 + 2(b_4 + 2(0)))))$$

Which, if we work it out step by step, becomes,

$$b_0 + 2(b_1 + 2(b_2 + 2(b_3 + 2(b_4 + 2(0)))))$$
$$= b_0 + 2(b_1 + 2(b_2 + 2(b_3 + 2b_4)))$$
$$= b_0 + 2(b_1 + 2(b_2 + 2b_3 + 2^2b_4))$$
$$= b_0 + 2(b_1 + 2b_2 + 2^2b_3 + 2^3b_4)$$
$$= b_0 + 2b_1 + 2^2b_2 + 2^3b_3 + 2^4b_4$$
$$= b_0 \times 2^0 + b_1 \times 2^1 + b_2 \times 2^2 + b_3 \times 2^3 + b_4 \times 2^4$$

which is the expanded form of a binary number with digits $b_4b_3b_2b_1b_0$.

Others to Decimal. The recurrence relation for converting binary numbers to decimal works for any number base. For example,

$$d_0 = 0$$
$$d_i = b_{k-i} + Bd_{i-1}, i = 1\ldots k$$

for a base B number with digits b_i.

Armed with this knowledge it becomes straightforward to convert hexadecimal or octal numbers to decimal. Simply use $B = 16$ for hexadecimal and $B = 8$ for octal. In the case of hexadecimal, which is a base greater than 10, the additional digits, $A - F$, take on their actual decimal values $10 - 15$.

As an example, let's convert $56AD_{16}$ to decimal. Since this is a four digit number $k = 4$ and the recurrence relation becomes,

$$d_0 = 0$$
$$d_i = b_{k-i} + 16d_{i-1}, i = 1 \ldots k$$

and calculation proceeds as,

i	d_i
0	0
1	$5 + 16(0) = 5$
2	$6 + 16(5) = 86$
3	$10 + 16(86) = 1386$ (*since* $A_{16} = 10$)
4	$13 + 16(1386) = 22189$ (*since* $D_{16} = 13$)

so $56AD_{16} = 22189$.

Finally, a similar calculation works for octal numbers. Convert 7502_8 to decimal. Again, $k = 4$ and the recurrence relation is now,

$$d_0 = 0$$
$$d_i = b_{k-i} + 8d_{i-1}, i = 1 \ldots k$$

giving,

i	d_i
0	0
1	$7 + 8(0) = 7$
2	$5 + 8(7) = 61$
3	$0 + 8(61) = 488$
4	$2 + 8(488) = 3906$

with a final result of $7502_8 = 3906$.

Frequent use of hexadecimal and octal numbers will quickly teach you to simply remember the first few powers of each base. Embedded systems programmers learn this early on for at least the first four digits. For hexadecimal, the powers of 16 are,

$$
\begin{array}{c|c}
16^0 & 1 \\
16^1 & 16 \\
16^2 & 256 \\
16^3 & 4096
\end{array}
$$

and similarly for octal,

$$
\begin{array}{c|c}
8^0 & 1 \\
8^1 & 8 \\
8^2 & 64 \\
8^3 & 512
\end{array}
$$

with these, small hexadecimal and octal integers can be converted to decimal quickly,

$$1234_{16} = 1 \times 4096 + 2 \times 256 + 3 \times 16 + 4 \times 1 = 4660$$
$$1234_8 = 1 \times 512 + 2 \times 64 + 3 \times 8 + 4 \times 1 \quad = 668$$

With a little practice, such conversion can be performed mentally.

1.4 Chapter Summary

In this chapter we briefly reviewed numbers, how they were represented historically, how place notation works, and how to manipulate numbers in the bases most frequently used by computers. It is helpful to become comfortable working with the conversion between number bases and it is especially helpful to memorize the binary representation of the first 16 integers. The exercises will help build this confidence and the skill will be used frequently in the remainder of this book.

Exercises

1.1. Using Egyptian notation and method calculate 22×48 and $713 \div 31$.

1.2. Using Babylonian notation calculate $405 + 768$. (*Hint:* carry on sixty, not ten)

1.3. Using Mayan notation calculate $419 + 2105$. (*Hint:* carry on twenty, not ten)

1.4. Convert the following hexadecimal numbers to binary: $A9C1_{16}$, $20ED_{16}$, $FD60_{16}$.

1.5. Convert the following octal numbers to binary: 773_8, 203_8, 656_8.

1.6. Using the table method, convert 6502 to binary.

1.7. Using the division method, convert 8008 to binary.

1.8. Write $e = 2.71828182845904\ldots$ to at least four decimal places using Babylonian notation. (*Hint:* You will need to use three negative powers of sixty) **

1.9. Write a program to convert a number in any base < 37 to its decimal equivalent using the recurrence relation. Use a string to hold the number itself and use $0 - 9$ and capital letters $A - Z$ for the digits. *

References

1. Boyer, C.B.: A History of Mathematics. Princeton University Press, Princeton, New Jersey (1985)
2. Gillings, R.: Mathematics in the Time of the Pharaohs. Dover Publications, New York (1982)
3. Menninger, K.: Number Words and Number Symbols. Dover Publications, New York (1992)

4. Neugebauer, O.: Mathematical Cuneiform Texts. American Oriental Series, vol. 29. New Haven: American Oriental Society (1945)
5. Neugebauer, O.: The Exact Sciences in Antiquity. Princeton University Press, Princeton, New Jersey (1952)

References

Chapter 2
Integers

Abstract Integers are perhaps the most important class of numbers. This is certainly true in the case of computers. In this chapter we dive into the integers and how computers represent and operate on them. Without these operations, digital computers would not function. We begin with some preliminary notation and terminology. Next we take a detailed look at the unsigned integers. We follow this with an examination of negative integers and their operations. Lastly, we finish with a look at binary coded decimals.

2.1 Bits, Nibbles, Bytes, and Words

Our tour of integers will be easier if we get some terminology out of the way first. This will make it just that much easier to talk about the way numbers are actually represented inside a computer. We'll start small and work our way up.

Bits. As we have seen, computers work with binary numbers. A single binary digit is known affectionately as a *bit*. This bit is either zero or one, off or on, false or true. The word *bit* is short for *binary digit* and appears in Claude Shannon's classic 1948 paper *A Mathematical Theory of Communication* [2]. Shannon attributed *bit* to John Tukey who used it in a Bell Labs memo dated January 9, 1947. Note, throughout this book we will play fast and loose with the bit terminology and freely alternate between "on" and "true" for when a bit is set to one and "off" or "false" when a bit is set to zero. The choice of word will depend on the context.

Since a bit is so simple, just a zero or a one, it is the smallest unit of information in the digital world. Still, this small unit can be very important. A status bit set to one may indicate that the rocket is ready to launch while a status bit set to zero indicates a fault. Indeed, some microcontroller devices, like the 8051, make excellent use of their limited resources and support single bit data. We will make extensive use

© Springer International Publishing Switzerland 2015

R.T. Kneusel, *Numbers and Computers*, DOI 10.1007/978-3-319-17260-6_2

of bits in the remainder of this chapter. Zero or one, how hard can it be? Still, as useful as a bit is, it is just, well, a *bit*. You need to group them together to get more expressive numbers.

Nibbles. A *nibble* (sometimes *nybble*) is a four bit number. As a four bit number a nibble can represent the first 16 integers with $0000_2 = 0$ and $1111_2 = 15$. And, as we now know, the numbers 0 through 15 are exactly the digits of a hexadecimal number. So, a nibble is a single hexadecimal digit. This makes sense since with four bits we can represent $2^4 = 16$ possible values. A nibble is also the amount of information necessary to represent a single binary-coded decimal digit but we are getting ahead of ourselves and will talk about binary-coded decimal later in this chapter. The origin of the word *nibble* is related to the origin of the word *byte* so we will expand on it below.

Modern computers do not use nibbles as a unit of data for processing. However, the very first microprocessors, like the TMS 1000 by Texas Instruments and the 4004 by Intel, both introduced in 1971, were 4-bit devices and operated on 4-bit data. This makes the nibble a bit like octal numbers: appearing in the shadows but not a real player in modern computing.

Bytes. For a modern computer, the smallest unit of data that it will work with is the *byte* which is an eight bit binary number. This means that the integers from $0 \ldots 255$ can be represented in a single byte. We'll ignore the concept of negative numbers for the time being.

The term *byte* was first used by Werner Buchholz in July 1956 during the design of the IBM Stretch computer and was an intentional misspelling of "bite" to avoid confusion with "bit" [3]. Since *byte* came from "bite" it is natural to be tempted to call half a byte, which is a four bit number, a "nibble" since a nibble is a small bite. Hence the origin of *nibble* above.

Buchholz intended a byte to represent a character. We still use bytes to represent characters in ASCII, but have since moved on to larger numbers for characters in Unicode. In this book, however, to keep things simple, we will make an implicit assumption that characters are bytes. So, a text that contains 1000 characters will take 1000 bytes of memory to store. The universal use of bytes when working with computers leads to the frequent appearance of the numbers $2^8 = 256$ and $2^8 - 1 = 255$. For example, colors on computers are often represented as triplets of numbers to indicate the amount of red, green, and blue that make up the color. Typically, these triplets are given as a set of three bytes to indicate the amount so that 000000_{16} is black and $00FF00_{16}$ is bright green. This is just one example, bytes are everywhere. The relationship between bits, nibbles and bytes is visualized in Fig. 2.1. Here we see that a byte can be thought of as two nibbles, each a hexadecimal digit, or as eight bits, each a binary digit.

Words. A byte is two nibbles, so, is a *word* two bytes? Well, sometimes, yes. And sometimes it is four bytes, or eight, or some other value. Unfortunately, the term *word* is not generic like bit and byte, instead it is tied to a particular computer architecture and describes the basic data size of the architecture.

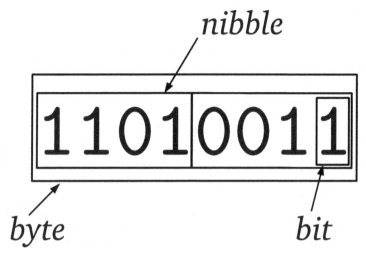

Fig. 2.1 A byte which consists of two nibbles and eight bits. The number is 211 which is made up of two nibbles, D_{16} and 3_{16} giving $D3_{16}$, or eight bits 11010011_2

Historically, the word size of a computer could be anywhere from four bits for early microprocessors (Intel 4004) to 60 bits for early mainframes (CDC 6600) but modern computers have settled on powers of two for their word sizes, typically 32 or 64 bits with some second generation microprocessors using 16 bit words (Intel 8086, WDC 65816). For our purposes, we can assume a 32 bit word size when describing data formats. On occasion we will need to be more careful and explicitly declare the word size we are working with.

With these preliminaries now under our belt, it is time to start working with actual data representations. We start naturally with unsigned integers.

2.2 Unsigned Integers

Unsigned integers are our first foray into the depths of the computer as far as numbers are concerned. These are the basic numbers, the positive integers, that find frequent use in representing characters, as counters, or as constants, really anything that can be mapped to the set $\{0, 1, 2, \ldots\}$. Of course, computers have finite memory so there is a limit to the size of an unsigned integer and we will get to that in the next section.

2.2.1 Representation

Integers are stored in memory in binary using one or more bytes depending on the range of the integer. The example in Fig. 2.1 is a one byte unsigned integer. If we are working in a typed language like C we need to declare a variable that will store one byte. Typically, this would mean using an *unsigned char* data type,

`unsigned char myByte;`

which can store a positive integer from $00000000_2 = 0$ to $11111111_2 = 255$. This is therefore the *range* of the `unsigned char` data type. Note how C betrays its age by referring to a byte number as a character. Unsigned integers larger than 255 require more than one byte.

Table 2.1 shows standard C types for unsigned integers and the allowed range for that type. These are fixed in the sense that numbers must fit into this many bits at all times. If not, an underflow or overflow will occur. Languages like Python abstract integers for the programmer and only support the concept of integer as opposed to a floating-point number. The unsigned integer operations discussed later in the chapter work nicely with Python but the concept of range is a little nebulous in that case since Python will move between internal representations as necessary.

Declaration	Minimum	Maximum	Number of bits
unsigned char	0	255	8
unsigned short	0	65,535	16
unsigned int	0	4,294,967,295	32
unsigned long	0	4,294,967,295	32
unsigned long long	0	18,446,744,073,709,551,615	64

Table 2.1 Unsigned integer declarations in C. The declaration, minimum value, maximum value and number of data bits are given

If a number like $11111111_2 = 255$ is the largest unsigned number that fits in a single byte how many bytes will be needed to store 256? If we move to the next number we see that we will need *nine* bits to store 256 which implies that we will need two bytes. However, there is a subtlety here that needs to be addressed. This is, if the number is to be stored in the computer's memory, say starting at address 1200, how should the individual bits be written to memory location 1200 and the one following (1201, since each memory location is a byte)? This question has more than one answer.

2.2.2 Storage in Memory: Endianness

Addresses. In order to talk about how we store unsigned integers in computer memory we have to first talk a bit about how memory is addressed. In this book, we have a working assumption of a 32-bit Intel-based computer running the Linux operating system. Additionally, we assume gcc to be our C compiler. In this case, we have *byte-addressable* memory meaning that even though the word size is four bytes, we can refer to memory addresses using bytes. For example, this small C program,

```
 1 #include <stdio.h>
 2
 3 int main() {
 4     unsigned char *p;
 5     unsigned short n = 256;
 6
 7     p = (unsigned char *)&n;
 8     printf("address of first byte  = %p\n", &p[0]);
 9     printf("address of second byte = %p\n", &p[1]);
10
11     return 0;
12 }
```

when compiled and run produces output similar to,

```
address of first byte  = 0xbfdafe9e
address of second byte = 0xbfdafe9f
```

where the specific addresses, given in hexadecimal and starting with the 0x prefix used by C, will vary from system to system and run to run. The key point is that the difference between the addresses is *one* byte. This is what is meant by a byte-addressable memory. If you don't know the particulars of the C language, don't worry. The program is defining a two byte number in line 5 (*n*) and a pointer to a single byte number in line 4 (*p*). The pointer stores a memory address which we set to the beginning of the memory used by *n* in line 7. We then ask the computer to print the numeric address of the memory location (line 8) and of the next byte (line 9) using the & operator and indexing for the first (p[0]) and second (p[1]) bytes. With this in mind, let's look at actually storing unsigned integers.

Bit Order. To store the number $11011101_2 = 221$ in memory we use the eight bits of the byte at a particular memory address. The question then becomes, which bits of 221 map to which bits of the memory? If we number the bits from 0 for the lowest-order bit, which is the right-most bit when writing the number in binary, to 7 for the highest-order bit, which is the left-most bit when writing in binary, we get a one-to-one mapping,

$$\frac{7\ 6\ 5\ 4\ 3\ 2\ 1\ 0}{1\ 1\ 0\ 1\ 1\ 1\ 0\ 1}$$

so that the highest-order bit of the number is put into the highest-order bit of the memory address. This seems sensible enough but one could imagine doing the reverse as well,

$$\frac{7\ 6\ 5\ 4\ 3\ 2\ 1\ 0}{1\ 0\ 1\ 1\ 1\ 0\ 1\ 1}$$

So, which is correct? The answer depends on how bits are read from the memory location. In modern computer systems, the bits are read low to high meaning the low-order bit is read first (bit 0) followed by the next higher bit (bit 1) so that the computer will use the first ordering of bits that maps 76543210 to 11011101.

Byte Order. Within a byte, bits are stored low to high. What about the bytes of a multibyte integer? We know that the number $100000000_2 = 256$ requires two bytes of memory, one storing the low set of eight bits 00000000_2 and another storing the high set of eight bits 00000001_2 where leading zeros are used to indicate that it is the first bit position set and the rest of the bits in the byte are clear (set to zero). But, what order in memory should we use,

Memory address:	1200	1201
Low first:	00000000	00000001
High first:	00000001	00000000

low first or high first? The answer is either. If we put the low order bytes of a multibyte number in memory first we are using *little-endian* storage. If we put the high order bytes first we are using *big-endian* or *network order*. The choice is the *endianness* of the number and both are in use typically as a function of the microprocessor, which has a preferred ordering. The terms little-endian and big-endian are derived from the book *Gulliver's Travels* by satirist Jonathan Swift [4]. Published in 1726 the book tells the story of Gulliver and his travels throughout a fictional world. When Gulliver comes to the land of Lilliput he encounters two religious sects who are at odds over whether to crack open their soft-boiled eggs little end first or big end first. Intel based computers use little-endian when storing multibyte integers. Motorola based computers use big-endian. Additionally, big-endian is called network order because the Internet Protocol standard [5] requires that numeric values in packet headers be stored in big-endian order.

Let's look at some examples of little and big endian numbers. To save space, we will use a 32-bit number but represent the numbers showing only the hexadecimal values of the four bytes such a number would occupy. For example, if the 32-bit number is,

$$10101110111100000101011010010110_2 = AE_{16} \, F0_{16} \, 56_{16} \, 96_{16}$$

we will write it as AEF05696 dropping the 16 base indicator for the time being. Putting this number into memory starting with byte address zero and using bigending ordering gives,

Address:	0	1	2	3
Value:	AE	F0	56	96

where the first byte one would read, pulling bytes from address zero first, is the high-order byte. If we want to use little-endian we would instead put the lowest order byte first to fill in memory with,

Address:	0	1	2	3
Value:	96	56	F0	AE

which will give us the low-order byte first when reading from address zero. Note, within a particular byte we always still use the bit ordering that maps the high-order bit (bit 7) to the high-order bit of the number. Reversing the bits within a byte when using little-endian is an error.

When we are working with data on a single computer within a single program, we typically do not need to pay attention to endianness. However, when transmitting data over the network or using data files written on other machines it is a good idea to consider what the endianness of the data is. For example, when reading sensor values from a CCD camera, which are typically 16-bit numbers requiring two bytes to store, it is essential to know the endianness otherwise the image generated from the raw data will look strange because the high-order bits which contain the majority of the information about the image will be in the wrong place. From the examples above, we see that correcting for endianness differences is straightforward, simply flip the order of the bytes in memory to convert from little-endian to big-endian and vice versa.

2.3 Operations on Unsigned Integers

Binary and unary operations on unsigned integers is the heart of what computers do. In this section we go over all the common operations, bit by bit, and mention some of the key things to look out for when working with unsigned integers of fixed bit width. Examples will use byte values to keep it simple but in some cases multibyte numbers will be shown. Specific operations will be shown in C and Python. Though the Python syntax is often the same as C, the results may differ because the Python interpreter will change the internal representation as necessary.

2.3.1 Bitwise Logical Operations

Bitwise operators are binary operators, meaning they operate on two numbers (called operands), to produce a new binary number. In the previous sentence the first instance of "binary" refers to the number of operands or arguments to the operation while the second instance of "binary" refers to the base of the operands themselves. This is an unfortunate but common abuse of notation.

Digital logic circuits, which are fascinating to study but well beyond the purview of this book, are built from logic gates which implement in hardware the logical operations we are discussing here. The basic set of logic operators are given names which relate to Boolean logic: AND, OR, XOR, and the unary NOT. Negated versions of AND and OR, called NAND and NOR, are also in use but these are easily constructed by taking the output of AND and OR and passing it through a NOT so we will ignore them here.

Logical operations are most easily understood by looking at their *truth tables* which illustrate the function of the operator by enumerating the possible set of inputs and giving the corresponding output. They are called truth tables because of the correspondence between the 1 and 0 of a bit and the logical concept of `true` and `false` which goes back to Boolean algebra developed by George Boole in the nineteenth century [6].

A truth table shows, for each row, the explicit inputs to the operator followed by the output for those inputs. For example, the truth table for the AND operator is,

AND		
0	0	0
0	1	0
1	0	0
1	1	1

which says that if the two inputs to the AND, recall the inputs are single bits, are 0 and 0 that the output bit will also be 0. Likewise, if the inputs are 1 and 1 the output will also be 1. In fact, for the AND operation the only way to get an output of 1 is for both input bits to be set. We can think of this as meaning "only both".

The two other most commonly used binary logic operators are OR and XOR. The latter is a called exclusive-OR and is sometimes indicated with EOR instead of XOR. Their truth tables are,

OR				XOR		
0	0	0		0	0	0
0	1	1		0	1	1
1	0	1		1	0	1
1	1	1		1	1	0

where we see that OR means "at least one" and XOR means "one or the other but not both". The last of the common bitwise logical operators is NOT and it has a very simple truth table,

NOT	
0	1
1	0

where the operation is simply turn 1 to 0 and 0 to 1.

If we look again at the truth tables for AND, OR and XOR we see that there are four rows in each table which match the four possible sets of inputs. Looking at these rows we can interpret the four bits in the output as a single four bit number. A four bit number has sixteen possible values from 0000_2 to 1111_2 which implies that there are sixteen possible truth tables for a binary bitwise operator. Three of them are AND, OR and XOR and another three are the tables made by negating the output of these operators. That leaves ten other possible truth tables. What do these correspond to and are they useful in computer programming? The full set of possible binary truth tables is given in Table 2.2 along with an interpretation of each operation. While all of these operators have a name and use in logic many are clearly of limited utility in terms of computer programming. For example, always outputting 0 or 1 regardless of the input (FALSE and TRUE in the table) is of no practical use in helping the programmer. The negated versions of the three main operators, NAND, NOR and XNOR, are naturally built by adding a "not" word to the positive versions and are therefore useful.

p:	0	0	1	1	
q:	0	1	0	1	
FALSE	0	0	0	0	always false
AND	0	0	0	1	p and q
	0	0	1	0	if p then not q else false
	0	0	1	1	p
	0	1	0	0	if not p and q
	0	1	0	1	q
XOR	0	1	1	0	p or q but not (p and q)
OR	0	1	1	1	p, q or (p and q)
NOR	1	0	0	0	not OR
XNOR	1	0	0	1	not XOR
	1	0	1	0	not q
	1	0	1	1	if q then p else true
	1	1	0	0	not p
	1	1	0	1	if p then q else true
NAND	1	1	1	0	not AND
TRUE	1	1	1	1	always true

Table 2.2 The sixteen possible binary bitwise logical operators. If a common name exists for the operation it is given in the first column. The top two rows labeled p and q refer to the operands or inputs to the operator. A 0 is considered false while 1 is true. A description of the operator is given in the last column. In the description the appearance of p or q in an if-statement implies that the operand is true

Now that we know the basic operators and why this set is most useful to computer programmers let's look at a few examples. First the AND operator,

$$
\begin{array}{rl}
& 10111101 = 189 \\
\text{AND} & 11011110 = 222 \\
\hline
& 10011100 = 156
\end{array}
$$

where the binary representation of the number is in the first column and the decimal is in the second. We see that for each bit in the two operands, 189 and 222, the output bit is on only if the corresponding bits in the operand were on. If the operands were simply zero or one the output would be one only if the inputs were both one. Since this is exactly what we mean when we use "and" in a sentence we can use this operator to decide if two things are both true. A natural place to use this ability is in an if-statement and many programming languages, C and Python included, use the concept that zero is false and one is true. Actually, C and Python extend this concept and declare anything that is not zero to be true. It should be noted that Python also supports Boolean variables and treats the keywords True and False as true and false respectively.

We look next at the OR and XOR operators. For OR we have,

```
      10111101 = 189
OR    11011110 = 222
      11111111 = 255
```

which means at in every bit position at least one of the operands had a bit set. The XOR operator will give us,

```
      10111101 = 189
XOR   11011110 = 222
      01100011 =  99
```

One useful property of the XOR operator is that it undoes itself if applied a second time. Above, we calculated 189 XOR 222 to get 99. If we now perform 99 XOR 222 we get,

```
      01100011 =  99
XOR   11011110 = 222
      10111101 = 189
```

giving us back what we started with. As an aside, this is useful as a simple way to encrypt data. If Alice wishes to encrypt a stream of n bytes (M), which may represent any data at all, so that she can send it (relatively) securely to Bob, all she need do is generate a stream of n random bytes (S) and apply XOR, byte by byte, to the original stream M to produce a new stream, M'. Alice can then transmit M' to Bob any way she wishes knowing that if Carol intercepts the message she cannot easily decode it without the same random stream S. Of course, Bob must somehow have S as well in order to recover the original message M. If S is truly random, used only once and then discarded, this is known as a *one-time pad* encryption. The keys are that S is truly random and that it is only used once. Of course, this does not cover the possibility that Carol may get her hands on S as well, but if she cannot, there is (relatively) little chance she will be able to recover M from just M'.

Another handy use for XOR is that it allows us to swap two unsigned integers without using a third variable. So, in C, instead of,

```
unsigned char a,b,t;
...
t = a;
a = b;
b = t;
```

we can use,

```
unsigned char a,b;
...
a ^= b;   // a = a ^ b;
b ^= a;
a ^= b;
```

where ^ is the C operator for XOR and a "= b is shorthand for a = a ^ b. Why does this work? If we write out an example in binary, we will see,

```
a = 01011101
b = 11011011
a ^= b      01011101 ^ 11011011 → a=10000110
b ^= a      10000110 ^ 11011011 → b=01011101
a ^= b      10000110 ^ 01011101 → a=11011011
```

As mentioned above, if we do a XOR $b \rightarrow c$ and then c XOR b we will get a back. If we instead do c XOR a we will get b back. So, the trick makes use of this by first doing a XOR b and then using that result, stored temporarily in a, to get a back but this time storing it in b and likewise for getting b back and storing it in a. This, then, swaps the two values in memory. This trick works for unsigned integers of any size. Another common use of XOR is in parity calculations. The parity of an integer is the number of 1 bits in its binary representation. If the number is odd, the parity is 1, otherwise it is 0. This can be used as a simple checksum when transmitting a series of bytes. A checksum is a indicator of the integrity of the data. If the checksum calculated on the receiving end does not match then there has been an error in transmission. If the last byte of data is the running XOR of all previous bytes the parity of this byte can be used to determine if a bit was changed during transmission. In order to see how to use XOR for parity calculations we need to wait a little bit until we talk about shifts below.

The effect of NOT is straightforward to illustrate,

$$\frac{\text{NOT}\ \ 10111101 = 189}{01000010 =\ \ 66}$$

it simply flips the bits from 0 to 1 and 1 to 0. We will see this operation again when we investigate signed integers.

Figure 2.2 gives a C program that implement the AND, OR, XOR and NOT logical operators. This illustrates the syntax. Note that the C example uses the unsigned char data type. This is an 8-bit unsigned integer like we saw in the examples above. The operators work with any size integer type so we could just as well have used unsigned short or unsigned int, etc. Figure 2.3 gives the corresponding Python code.

The standard C language does not support direct output of numbers in binary so the example in Fig. 2.2 instead outputs in decimal and hexadecimal. Recall that in Chap. 1 we learned how to easily change hexadecimal numbers into binary by replacing each hexadecimal digit with four bits representing that digit. The Python code makes use of standard string formatting language features to output in binary directly. As Python is itself written in C and the language was designed to show that heritage, the syntax for the bitwise logical operators is the same. There is one subtlety with the Python code you may have noticed. The function pp() uses "z & 0xff" in the format() method call where you might have expected only "z". If we do not AND the output value with FF_{16} we will output a full range integer

```
1 #include <stdio.h>
2
3 void pp(unsigned char z) {
4     printf("%3d (%02x)\n", z, z);
5 }
6
7 int main() {
8     unsigned char z;
9
10     z = 189 & 222;  pp(z); // '&' is AND
11     z = 189 | 222;  pp(z); // '|' is OR
12     z = 189 ^ 222;  pp(z); // '^' is XOR
13     z = z ^ 222;    pp(z); // returns 189
14     z = ~z;         pp(z); // '~' is NOT
15
16     return 0;
17 }
```

Fig. 2.2 Bitwise logical operators in C

```
1 def pp(z):
2     print "{0:08b}".format(z & 0xff)
3
4 def main():
5     z = 189 & 222;  pp(z) # '&' is AND
6     z = 189 | 222;  pp(z) # '|' is OR
7     z = 189 ^ 222;  pp(z) # '^' is XOR
8     z = z ^ 222;    pp(z) # returns 189
9     z = ~z;         pp(z) # '~' is NOT
10
11 main()
```

Fig. 2.3 Bitwise logical operators in Python

and as we will see later in this chapter, that would be a negative number to Python. The AND keeps only the lowest eight bits and gives us the output we would expect matching the C code in Fig. 2.2.

2.3.2 Testing, Setting, Clearing, and Toggling Bits

Common operations on unsigned integers include the setting, clearing and testing of particular bits. Setting a bit means to turn that bit on (1) while clearing is to turn the bit off (0). Testing a bit simply returns its current state, 0 or 1, without changing its value. In the embedded computing world this is often done to set an output line high or low or to read the state of an input line. For example, a microcontroller uses specific pins on the device as digital inputs or outputs. Typically, this requires setting bits in a control register or reading bits in a register to know whether a voltage is present or not on pins of the device.

The bitwise logical operators introduced above are the key to working with bits. As Fig. 2.3 already alluded to, the AND operator can be used to mask bits. Masking sets certain bits to zero and some examples will make this clear. First, consider a situation similar to the Python code of Fig. 2.3 that keeps the lower nibble of a byte while it clears the upper nibble. To do this, we AND the byte with $0F_{16}$,

$$
\begin{array}{r}
10111101 = BD \\
AND \quad 00001111 = 0F \\
\hline
00001101 = 0D
\end{array}
$$

where we now display the binary and hexadecimal equivalent. The output byte will have all zeros in the upper nibble because AND is only true when both operands are one. Since the second operand has all zeros in the upper nibble there is no situation where the output can be one. Likewise, the lower nibble is unaffected by the AND operation since every position where the first operand has a one the second operand will also have a one leading to an output of one but every position where the first operand has a zero the second operand will still have a one leading to an output of zero, which is just what the first operand has in that position. A quick look back at the truth table for AND (above) will clarify why this masking works.

This property of AND can be used to test if a bit is set. Say we wish to see if bit 3 of BD_{16} is set. All we need to do is create a mask that has a one in bit position 3 and zeros in all other bit positions (recall that bit positions count from zero, right to left),

$$
\begin{array}{r}
10111101 = BD \\
AND \quad 00001000 = 08 \\
\hline
00001000 = 08
\end{array}
$$

if the result of the AND is not zero we know that the bit in position 3 of BD_{16} is indeed set. If the bit was clear, the output of the entire operation would be exactly zero which would indicate that the bit was clear. Since C and Python treat nonzero as true the following C code would output "bit 3 on",

```
if (0xBD & 0x08) {
    printf("bit 3 on\n");
}
```

as would this Python code,

```
if (0xBD & 0x08):
    print "bit 3 on"
```

We use AND to test bits and mask bits but we use OR to actually set bits. For example, to actually set bit 3 and leave all other bits as they were we would do something like this,

$$
\begin{array}{r}
10110101 = B5 \\
OR \quad 00001000 = 08 \\
\hline
10111101 = BD
\end{array}
$$

which works because OR returns true when either operand or both have a one in a particular bit position. Just as AND with an operand of all ones will return the other operand, so an OR with all zeros will do likewise. In order to set bits, then, we need OR, a bit mask, and assignment as this C code shows,

```
#include <stdio.h>

int main() {
    unsigned char z = 0xB5;

    z = z | 0x08;   // set bit 3
    printf("%x\n", z);
    z |= 0x40;      // set bit 6
    printf("%x\n", z);

    return 0;
}
```

which outputs BD_{16} and FD_{16} after setting bits 3 and 6. Since we updated z each time the set operations are cumulative.

We've seen how to test and set bits now let's look at clearing bits without affecting other bits. We know that if we AND with a mask of all ones we will get back our first operand. So, if we AND with a mask of all ones that has a zero in the bit position we want to clear we will turn off that bit and leave all other bits as they were. For example, to turn off bit 3 we do something like this,

$$
\begin{array}{rl}
 & 10111101 = BD \\
AND & 11110111 = F7 \\
\hline
 & 10110101 = B5
\end{array}
$$

where the mask value for AND is $F7_{16}$. But, this begs the question of how we get the magic $F7_{16}$ in the first place? This is where NOT comes into play. We know that to set bit 3 we need a mask with only bit 3 on and all others set to zero. This mask value is easy to calculate because the bit positions are simply powers of two so that the third position is $2^3 = 8$ implying that the mask is just $8 = 00001000_2$. If we apply NOT to this mask we will invert all the bits giving us the mask we need: 11110111_2. This is readily accomplished in C,

```
unsigned char z = 0xBD;
z = z & (~0x08);
printf("%x\n", z);
```

which outputs $B5_{16} = 10110101_2$ with bit 3 clear. The same syntax will also work in Python.

Our last example toggles bits. When a bit is toggled its value is changed from off to on or on to off. We can use NOT to quickly toggle all the bits, since this is what NOT does, but how to toggle only one bit and leave the rest unchanged? The answer lies in using XOR. The XOR operator returns true only when the operands are different. If both are zero or one, the output is zero. This is what we need. If we XOR our value with a mask that has a one in the bit position we desire to toggle we will get something like this for bit 1,

$$
\begin{array}{r}
10111101 = \text{BD} \\
\text{XOR} \quad 00000010 = 02 \\
\hline
10111111 = \text{BF}
\end{array}
$$

where we have turned bit 1 on when previously it was off. If bit 1 was already on we would have 1 XOR 1 = 0 which would turn it off in the output. Clearly, we can toggle multiple bits by setting the positions we want to toggle in the mask and then apply the XOR. In C we have,

```
unsigned char z = 0xBD;
z = z ^ 0x02;
printf("%x\n", z);
```

which will output $\text{BF}_{16} = 10111111_2$ as expected.

2.3.3 Shifts and Rotates

So far we have looked at operations that manipulate bits more or less independently of other bits. Now we take a look at sliding bits from one position to another within the same value. These manipulations are accomplished through the shift and rotate operators. A *shift* is as straightforward as it sounds, just move bits from lower positions in the value to higher, if shifting to the left, or from higher positions to lower if shifting to the right. When shifting, bits simply "fall off" the left or right if they hit the end of the integer. This implies something, namely, that we impose a specific number of bits on the integer. For our examples we will stick with 8-bit unsigned integers though all of these operations work equally well on integers of any size. Let's look at what happens when we shift a value to the left one bit position using binary notation,

$$10101111 \leftarrow 1 = 01011110$$

where we use the $\leftarrow$ symbol to mean shift to the left and the 1 is the number of bit positions. The leading 1 drops off the left end and a zero moves in from the right end to fill in the empty space. All other bits move up one bit position. Now, what happens when only a single bit is set and we shift one position to the left,

$$00000010 \leftarrow 1 = 00000100$$

we see that we started with a value of $2^1 = 2$ and we ended with a value of $2^2 = 4$. Therefore, a single position shift to the left will move each bit to the next highest bit position which is the same as multiplying it by two. Since this will happen to all bits, the net effect is to multiply the number by two. Of course, if bits that were set fall off the left end we will lose precision but the remaining bits will be multiplied by two. For example,

$$00101110 \leftarrow 1 = 01011100$$

which takes $00101110_2 = 46$ to $01011100_2 = 92$ which is 46×2. Shifting to the left by more than one bit position is equivalent to repeated shifts by one position so shifting by two positions will multiply the number by $2 \times 2 = 4$,

$$00101110 \leftarrow 2 = 10111000$$

giving $10111000_2 = 184$ as expected.

It is natural to think that if a left shift multiplies by two a right shift would divide by two and this is indeed the case,

$$00101110 \rightarrow 1 = 00010111$$

gives $00010111_2 = 23$ which is $46 \div 2$. Just as bits falling off the left end of the number will result in a loss of precision so will bits falling off the right end with one significant difference. If a bit falls off the right end of the number it is lost from the ones position. If the ones position is set, that means the number is odd and not evenly divisible by two. If the bit is lost the result is still the number divided by two but the division is *integer division* which ignores any remainder. Information is lost since a right shift followed by a left shift results in a number that is one less than what we started with if the original number was odd. We can see this by shifting two positions to the right,

$$00101110 \rightarrow 2 = 00001011$$

to get $00001011_2 = 11$ which is $46 \div 4$ ignoring the remainder of 2.

Both C and Python support shifts using the $<<$ and $>>$ operators for left and right shifts respectively. The $<<$ operator is frequently used with an argument of 1 in order to quickly build bit masks,

$$1 << 3 = 00001000_2$$
$$1 << 5 = 00100000_2$$

which is a pretty handy way to move bits into position.

Before we leave shifts, let's return to the parity calculation mentioned above. Recall that the parity of a number is determined by the number of 1 bits in its binary representation. If odd, the parity is 1, otherwise it is 0. The key observation here is that XOR preserves the parity of its arguments. For example, if, in binary, a = 1101 and b = 0111 then the parity of the two together is zero since there are a total of six on bits. If we apply XOR we get a XOR b = 1010 which also has an even number of on bits and therefore has the same parity. This suggests the trick. If we XOR a number with itself but first shift the number half its bit width the resulting bits of the lower half of the output of XOR will have the same parity as the original number. We can see this if we look at a = 01011101 and XOR it with itself after shifting down by four bits, which is half the width of the number,

$$
\begin{array}{r}
01011101 \\
\text{XOR } 00000101 \\
\hline
\text{xxxx}1000
\end{array}
$$

where we are ignoring the upper four bits. The original number has 5 on bits therefore the parity is 1. If we look at the lower four bits after the XOR we see it also has a parity of 1. If we repeat the process but using the new value and this time shifting by half its effective width, which is 4 bits, we will end up with a number that has the same parity as we started with and the same parity as the original number. If we repeat this all the way the final result will be a one bit number that is the parity of the original number. Therefore, for 01011101,

original	01011101
right shift 4	00000101
XOR	01011000
right shift 2	00010110
XOR	01001110
right shift 1	00100111
XOR	01101001
parity bit	xxxxxxx1

For an 8-bit number we can define a parity function in C as,

```
unsigned char parity(unsigned char x) {
    x = (x >> 4) ^ x;
    x = (x >> 2) ^ x;
    x = (x >> 1) ^ x;

    return x & 1;
}
```

where the `return x & 1;` line returns only the lowest order bit since all other bits are meaningless at this point. The extension of this function to wider integers is straightforward.

Shifting moves bits left or right and drops them at the ends. This is not the only option. Instead of dropping the bits we might want to move the bits that would have fallen off to the opposite end. This is a *rotation* which like a shift can be to the left or right. Unfortunately, while many microprocessors contain rotate instructions as primitive machine language operations neither C nor Python support rotations directly. However, they can be simulated easily enough in code and an intelligent compiler will even turn the simulation into a single machine instruction (e.g., gcc).

Again, to keep things simple, we will work with 8-bit numbers. For an 8-bit number, say $AB_{16} = 10101011_2$, a rotation to the right of one bit looks like this,

$$10101011 \Rightarrow 1 = 11010101$$

where we introduce the $\Rightarrow$ symbol to mean rotation to the right instead of simple shifting ($\rightarrow$). Rotation to the left bit position gives,

$$10101011 \Leftarrow 1 = 01010111$$

where we see that the bit that would have fallen off at the left end has now moved around to the right side.

To simulate rotation operators in C and Python we use a combination of << and >> with an OR operator. We need to pay attention to the number of bits in the data type in this case. For example, in C, we can define rotations to the right by any number of bits using,

```
unsigned char rotr(unsigned char x, int n) {
    return (x >> n) | (x << 8 - n);
}
```

with a similar definition for rotations to the left,

```
unsigned char rotl(unsigned char x, int n) {
    return (x << n) | (x >> 8 - n);
}
```

where we have changed << to >> and vice versa. One thing to note is the 8 in the second line of these functions. This number represents the number of bits used to store the data value. Since we have declared the argument x to be of type unsigned char we know it uses eight bits. If we wanted to modify the functions for 32-bit integers we would replace the 8 with 32 or, in general, use sizeof(x) * 8 to convert the byte size to bits.

The Python versions of the C rotation functions are similar.

```
def rotr(x,s):
    return ((x >> s) | (x << 8 - s)) & 0xff
```

and,

```
def rotl(x,s):
    return ((x << s) | (x >> 8 - s)) & 0xff
```

where the only change is that the return value is AND'ed with FF_{16} which as we have seen will keep the lowest eight bits and set all the others to zero. This is again because Python uses 32-bit integers internally and we are interested in keeping the output in the 8-bit range. If we wanted these functions to work with 16-bit integers, we would replace the 8 with 16, as in the C versions, but we would also need to make the mask keep the lowest 16 bits by replacing 0xff with 0xffff.

The rotation functions are helpful, but why do they work? Let's take a look by breaking the operations up individually and seeing how they combine to produce the final result. We start with the original input, $AB_{16} = 10101011_2$, and show, row by row, the first shift to the left, then the second shift to the right, and finally the OR to combine them. The values between the vertical lines are those that fit within the 8-bit range of the number, the other bits are those that are lost in the shift operations,

$$
\begin{array}{rl|l|l}
        & |10101011| & AB_{16} \\
        & |01010101|1 & AB_{16} >> 1 \\
1010101 & |10000000| & AB_{16} << 8\text{-}1 \\
        & |11010101| & OR
\end{array}
$$

where we see that the bit lost when shifting to the right one position has been added back at the front of the number by the shift to the left and the OR operation. Since the shifts always introduce a zero bit the OR will always set the output bit to the proper value because 1 OR 0 = 1 and 0 OR 0 = 0. This works for any number of bits to rotate,

$$
\begin{array}{ll}
10101011 & AB_{16} \\
00101010\,|11 & AB_{16} >> 2 \\
1010101\,|11000000 & AB_{16} << 8\text{-}2 \\
11101010 & OR
\end{array}
$$

2.3.4 Comparisons

Magnitude comparison operators take two unsigned integers and return a truth value about whether or not the relationship implied by the operator holds for the operands. Here we look at the basic three, equality ($A = B$), greater than ($A > B$), and less than ($A < B$). There is a second set which can be created easily from the first: not equal ($A \neq B$), greater than or equal ($A \geq B$), and less than or equal ($A \leq B$).

At its core, a computer uses digital logic circuits to compare bits. Figure 2.4 illustrates the layout of a 1-bit comparator. Comparators for larger numbers of bits are repeated instances of this basic pattern. As this not a book on digital logic we will go no further down this avenue but will instead talk about comparing unsigned integers from a more abstract point of view.

Fig. 2.4 A 1-bit digital comparator. The three output values represent the three possible relationships between the input values A and B. Recall that input values are 0 or 1. The output is 1 for the relationship that is true and 0 for those that are not meet. Cascades of this basic form can be used to create multi-bit comparators

Most microprocessors have primitive instructions for comparison of integers as this operation is so fundamental to computing. In addition to direct comparison, many instructions affect processor flags based on privileged numbers like zero. For example, to keep things simple, the 8-bit 6502 microprocessor, which has a single accumulator, A, for arithmetic, performs comparisons with the CMP instruction but also sets processor status flags whenever a value is loaded from memory using the LDA instruction. There are other registers and instructions, of course, but we focus on the accumulator to keep the example simple. The 6502 uses branch instructions like BEQ and BNE to branch on equal or not equal respectively. This also applies to the implied comparison with the special value 0 which happens when the LDA instruction loads the accumulator.

Armed with this brief review of an old 8-bit microprocessor we can see that the following set of instructions would indeed perform an 8-bit comparison between 22_{16} already in the accumulator via the first LDA instruction and $1A_{16}$ stored in memory location 2035_{16} and branch if they are not equal. Additionally, we will also perform an implicit comparison of the value of memory location $20FE_{16}$ with zero and branch if it is,

```
LDA #$22       ; A = $22
CMP $2035      ; compare to location $2035
BNE noteql     ; branch to "noteql" if not equal

LDA #$20FE     ; A = contents of $20FE
BEQ iszero     ; branch to "iszero" if zero
```

where we use the classic notation of $\$22 = 22_{16}$.

Why bring up this example? In part to show that comparison is very fundamental to computers and is done as efficiently as possible in hardware and to set the stage for our less efficient alternatives to digital logic. The comparisons we are implementing in code are pure hardware even in the simplest of microprocessors.

Since for any two integers A and B exactly one of the following is true: $A = B$, $A < B$, or $A > B$, it follows that if we know how to test for any two the last condition is simply when neither of the two we can test for applies. In our case, we look at the situation where we know how to test for equality ($A = B$) and greater than ($A > B$). We do this with two predicate functions that simply return 1 if the relationship holds for the arguments and 0 otherwise. Let's call these predicates isZero(A) and isGreater(A,B) and see how we might implement them directly in C for 8-bit values using the unsigned char data type.

You may be wondering why we chose to use isZero(A) instead of the perhaps more obvious isEqual(A,B). If so, good, you are paying attention. Given our experience with the XOR operator we now know that,

$$a \text{ XOR } a \to 0$$

so we can immediately see that,

$$isEqual(A,B) = isZero(A \text{ XOR } B)$$

but how do we implement isZero(A)? One approach in code would be to shift the bits and test the lowest order one. If we find one that is not zero then the number is not zero. The test is via an OR which is only zero when both operands are zero. We use AND to do the bit comparison and then shift the result down so that the compared bit is in the lowest position. Then, the OR of all these tests will be 1 if any bits are set and 0 if not. This is the exact opposite of what we want so we add a NOT to reverse the sense of the logic and a final AND with 1 to mask out all other bits

and return the state of the lowest bit only. This, then, is the full predicate function
`isZero(x)`,

```
unsigned char isZero(unsigned char x) {
    unsigned char ans;

    return ~(((x & (1<<7)) >> 7) |   // test bit 7
             ((x & (1<<6)) >> 6) |
             ((x & (1<<5)) >> 5) |
             ((x & (1<<4)) >> 4) |
             ((x & (1<<3)) >> 3) |
             ((x & (1<<2)) >> 2) |
             ((x & (1<<1)) >> 1) |
              (x & 1)) & 1;          // test bit 1
}
```

Notice that there are no actual comparison operators in this function, only logical
bitwise operators. With this we can quickly implement `isEqual(x,y)`,

```
unsigned char isEqual(unsigned char x,
                      unsigned char y) {
    return isZero(x ^ y);
}
```

We now need `isGreater(A,B)` which is implemented with bit operators,
shifts, and a call to `isZero(x)`. This is why we started with `isZero(x)` instead of `isEqual(x,y)`. The C code for our function is given first followed by
the explanation of why it works,

```
 1 unsigned char isGreater(unsigned char a,
 2                         unsigned char b) {
 3     unsigned char x,y;
 4
 5     x = ~a & b;
 6     y = a & ~b;
 7
 8     x = x | (x >> 1);
 9     x = x | (x >> 2);
10     x = x | (x >> 4);
11
12     return ~isZero(~x & y) & 1;
12 }
```

In order to tell if a is greater than b we need to know the first place where their
respective bits do not agree. Once we know this bit position we know that a is
greater than b if at that bit position a has a 1 while b has a 0. So, we need to find
the locations of where the bits differ. To make the example concrete, we let a be
$00011101_2 = 29$ and b be $00011010_2 = 26$. If we look at line 5 we see,

$$x = ~a \& b;$$

which sets x to the AND of the NOT of a and b. This leaves x with a 1 in all the
positions where the bit in a is less than the same bit in b. For our example this

sets x to 00000010_2 which tells us that the only bit position in a that is less than the corresponding bit position in b is bit 1. Likewise, line 6 asks where are the bit positions where a is greater than b? In this case, we set y to 00000101_2 to indicate that in bit 0 and bit 2 the value of a is greater than b. In order to see if a is greater than b we need to find the highest bit position where the two differ and see if that bit is set. We can do this if we take the value in x, which tells us where a bits are less than b bits, and build a mask which is 1 for all bit positions at or below the highest bit position where a is first less than b. We do this with lines 8 through 10. This operation which ORs the value with shifted versions of itself duplicates the highest 1 bit among all the lower bits. In this case,

```
x                        → 00000010
x = x | (x >> 1) → 00000011
x = x | (x >> 2) → 00000011
x = x | (x >> 4) → 00000011
```

where the last two steps add nothing new since x >> 2 and x >> 4 both result in zero which will set no new bits. We now have a mask in x that tells us all the bit positions below the first place where the bit in a is less than the bit in b. If we NOT this mask, ¬ 00000011 → 11111100, we can use the new mask to tell us all the bit positions where a was not less than b. Lastly, with this mask and the value in y which tells us where the bits in a were greater than the bits in b, we can perform one final AND, ~x & y, which will result in zero if a ≤ b since no bits will be set in y in the region where the bits of a were greater than those of b, or a nonzero value since at least one bit will be set in y in that region. Line 12, then, asks if this result is zero by passing the output to isZero. It then applies NOT to change the output of isZero since the result is zero when a ≤ b and not zero when a > b. The final AND with 1 gives us only the final bit since the NOT will change all the bits of the result of isZero.

We are nearly finished with our comparison operators. We have equality (isEqual) and greater than (isGreater). With these we see that isLess would be,

```
unsigned char isLess(unsigned char x,
                     unsigned char y) {
    return (!isEqual(x,y)) && (!isGreater(x,y));
}
```

which is reasonable since for any two unsigned integers A and B, if $A \ngtr B$ then $A < B$ must be true. Testing for not equal is even simpler,

```
unsigned char isNotEqual(unsigned char x,
                         unsigned char y) {
    return !isEqual(x,y);
}
```

since the only way for A to *not* not equal B is if the two are indeed equal. Less than or equal and greater than or equal follow directly from the functions already defined,

```
unsigned char isLessOrEqual(unsigned char x,
                            unsigned char y) {
    return isEqual(x,y) || isLess(x,y);
}
```

and

```
unsigned char isGreaterOrEqual(unsigned char x,
                               unsigned char y) {
    return isEqual(x,y) || isGreater(x,y);
}
```

which completes our implementation of comparison operators using only bitwise operations.

2.3.5 Arithmetic

Just as comparison is a fundamental microprocessor operation, so is arithmetic. In this section we look at arithmetic with unsigned binary integers, first from the point of view of doing it "by hand" and then from the point of view of a simple 8-bit microprocessor. These two approaches should illustrate the mechanism behind the operations. We will not, however, attempt to implement these operations in C as we did above for the comparison operators though we will use C to demonstrate overflow and underflow conditions.

The addition facts in binary are,

$$
\begin{aligned}
0 + 0 &= 0 \\
0 + 1 &= 1 \\
1 + 0 &= 1 \\
1 + 1 &= 0 \; \text{carry} \; 1
\end{aligned}
$$

from which we see that one of them produces a carry since it results in a two digit number. Just as in decimal, the carry is applied to the next digit over to the left. So, to add two unsigned 8-bit binary numbers we move right to left, bit by bit, adding and moving any carry to the next digit to the left,

$$
\begin{array}{r}
11111 \quad \leftarrow \textit{carry} \\
01101110 \leftarrow \textit{first operand} \\
+ \; 00110101 \leftarrow \textit{second operand} \\
\hline
10100011 \leftarrow \textit{answer}
\end{array}
$$

where the carry from the second to leftmost bit does not cause difficulty since the highest bits of each number are zero. However, what would happen if there was a carry from the leftmost bit? In mathematics, nothing special would happen, there would simply be a new leftmost bit, but in computers this is not the case. Recall that we are working with 8-bit unsigned numbers which means that all numbers fit in eight bits in memory. If we use eight bits for numbers we have no place in which to put any final carry bit. This results in an *overflow* condition. The computer simply discards this new highest value bit and retains the lowest eight bits which fit in memory,

$$
\begin{array}{r}
11111 \\
11101110 \quad EE_{16} \\
+ \quad 00110101 \quad 35_{16} \\
\hline
1 \;\; 00100011 \quad 123_{16}
\end{array}
$$

which is stored as 23_{16} discarding the carry on the leftmost bit. This is precisely what we see with the following C code,

```
#include <stdio.h>

int main() {
    unsigned char x, y, z;

    x = 0xEE;
    y = 0x35;
    z = x + y;

    printf("%x\n", z);
}
```

Let's take a look at how a simple 8-bit microprocessor would implement an addition. In this case, we look at an unsigned 16-bit addition which requires two addition operations. Working again with the 6502 processor mentioned above we see that an 8-bit addition between a value in memory locations 23_{16} and 25_{16} will involve a load into the accumulator (LDA), a clearing of the carry flag which catches any overflow bit (CLC) and an addition with memory (ADC). Specifically, we assume memory location 23_{16} contains EE_{16}, memory location 25_{16} contains 35_{16}. We then load the accumulator, clear the carry and add,

$$
\begin{array}{l|l}
\text{LDA \$23} & A \leftarrow EE \\
\text{CLC} & C \leftarrow 0 \\
\text{ADC \$25} & A \leftarrow A + 35 + C \\
& A \leftarrow 23, C \leftarrow 1
\end{array}
$$

with the accumulator set to 23_{16}, the lowest eight bits of the sum of EE_{16} and 35_{16}, and the carry flag set to 1 to indicate an overflow happened. This setting of the carry flag is the key to implementing multibyte addition. In C, we would simply declare the variables to be of data type unsigned short which is 16-bits and add as before. For example, in C we have,

```
#include <stdio.h>

int main() {
    unsigned short x, y, z;

    x = 0xEE;
    y = 0x35;
    z = x + y;

    printf("%x\n", z);
}
```

which gives us a 16-bit answer of 123_{16}. In memory, using little-endian representation for multibyte numbers, we have,

$$\text{memory location } \$24 : 23$$
$$\$25 : 01$$

since we store the lowest byte first.

In the simpler world of the 8-bit microprocessor we store the lowest part of the sum, the low byte, and add the high bytes without clearing the carry. Assuming memory is set to

$$\text{memory location } \$23 : EE$$
$$\$24 : 00$$
$$\$25 : 35$$
$$\$26 : 00$$

we clear the carry flag, add the two low bytes, store the partial sum, add the high bytes with any carry, and store the final part of the sum like this,

LDA $23	$A \leftarrow EE$
CLC	$C \leftarrow 0$
ADC $25	$A \leftarrow A + 35 + C, \ A{=}23$
STA $27	$\$27 \leftarrow 23, C \leftarrow 1$
LDA $24	$A \leftarrow 0$
ADC $26	$A \leftarrow A + 0 + C, \ A{=}1$
STA $28	$\$28 \leftarrow 1$

where we have introduced a new instruction, STA, which stores the accumulator in memory. When this sequence of instructions is complete we have the final answer in memory locations $27 and $28 as 23_{16} and 1_{16} respectively as we expect for a little-endian number.

The addition above is equivalent to this single 16-bit addition

$$
\begin{array}{r}
1 \ \ 11111 \\
00000000 \ \ 11101110 \quad EE_{16} \\
+ \quad 00000000 \ \ 00110101 \quad 35_{16} \\
\hline
00000001 \ \ 00100011 \quad 123_{16}
\end{array}
$$

where we have separated the upper eight bits of the high byte from the lower eight bits of the low byte.

Addition of unsigned binary numbers is by now straightforward. We add, left to right, bit by bit with carry bit when necessary. If the result is too large, we overflow and retain the lowest n bits where n is the width of the number in bits. We now move on to subtraction of unsigned binary numbers.

The subtraction facts in binary are,

$$0 - 0 = 0$$
$$0 - 1 = 1, \ \textit{underflow}$$
$$1 - 0 = 1$$
$$1 - 1 = 0$$

where the *underflow* condition will require a borrow from the next higher bit position. Like overflow, *underflow* is the situation where we cannot properly represent the number using the number of bits we have to work with. In this case, the underflow happens when we attempt to subtract a larger number from a smaller and we have no way to represent the resulting negative number. We'll address this issue below.

To continue with the example we used for addition, we now evaluate $EE_{16} - 35_{16}$ to get $B9_{16}$. In binary, using the subtraction facts, we have,

```
      01
      01   01
   11101110  EE16
 − 00110101  3516
   10111001  B916
```

with each borrow written above the bit and the next bit set to one less than it was. If a second borrow is necessary for a bit position, we write it above again. Let's look at the subtraction again, bit by bit, right to left, we are computing,

bit 0	10 - 1 = 1, *borrow*
bit 1	0 - 0 = 0
bit 2	1 - 1 = 0
bit 3	1 - 0 = 1
bit 4	10 - 1 = 1, *borrow*
bit 5	10 - 1 = 1, *borrow*
bit 6	0 - 0 = 0
bit 7	1 - 0 = 1

which, reading from bottom to top, gives $10111001_2 = B9_{16}$ as expected.

What happens if we need to borrow across more than one bit position? For example, in base 10 a problem like,

```
   7003
 -  972
   6031
```

involves borrowing across two digits in order to subtract the 7 of 972 which we can write as,

```
   69103
 -  9 72
   6031
```

where we change the 700 into 69^10 to subtract 9 7 giving the partial result 603. We subtracted one from the next two digits and added it in as a ten to the digit we were working with. The same thing happens in binary. Consider this subtraction problem,

$$
\begin{array}{r}
\texttt{01} \\
\texttt{1010}\underline{\texttt{10}}^{\texttt{1}}\texttt{01} \quad \texttt{A9}_{16} \\
-\quad \texttt{101001 10} \quad \texttt{A6}_{16} \\
\hline
\texttt{00000011} \quad \texttt{03}_{16}
\end{array}
$$

where we attempt to subtract 1 from 0 in the second bit position (bit 1, since we always count from zero and from right to left). We need to borrow from bit 2, but since this is also zero, we instead borrow from bit 3 and change $\texttt{100}$ into $\texttt{01}^{\texttt{1}}\texttt{0}$ in order to do the subtraction.

As we are working with 8-bit unsigned integers one will eventually be tempted to ask what happens if we try to subtract 1 from 0 since we cannot represent -1. What happens is we get the result we would get if we had an extra bit at left-most position and borrowed from it like so,

$$
\begin{array}{r}
\texttt{0 1111111} \\
\underline{\texttt{1}}\ \underline{\texttt{0000000}}^{\texttt{1}}\texttt{0} \\
-\quad \texttt{0000000 1} \\
\hline
\texttt{11111111}
\end{array}
$$

meaning that subtracting one from the smallest number we can represent, namely zero, results in the largest number we can represent which for an 8-bit unsigned integer is every bit set, $2^8 - 1 = 255$. Another way to think about it is that the numbers form a loop with $\texttt{00000000}$ and $\texttt{11111111}$ set next to each other. If we move one position down below zero we wrap around to the top and get $\texttt{255}$. Likewise, if we move one position up from $\texttt{255}$ we will wrap around to the bottom and get zero. Moving down below zero is an underflow condition while moving up above $\texttt{255}$ is an overflow.

Before we move on to multiplication and division of unsigned integers, let's look at one more subtraction example that would result in a negative number. We'll use our running example but swap the operands,

$$
\begin{array}{r}
\texttt{1 00110101} \quad \texttt{35}_{16} \\
-\quad \texttt{11101110} \quad \texttt{EE}_{16} \\
\hline
\texttt{01000111} \quad \texttt{47}_{16}
\end{array}
$$

where we have indicated the implied 1 bit from which we can borrow. This implied 1 bit is in the bit position for $2^8 = 256$ which suggests another way to think about the answer we would expect if we go negative in a subtraction of unsigned numbers. For our example, $\texttt{35}_{16} - \texttt{EE}_{16} = -185$, but if we add in 256 which is essentially what we are doing in thinking there is an implied extra high bit, we get $-185 + 256 = 71 = \texttt{47}_{16}$ which is the answer we found previously. We have been working with 8-bit wide unsigned integers. If we are using 16-bit integers the implied bit 16 (recall, we count bits from zero) is $2^{16} = 65536$ which means we would add 65536 to any negative value to get the result we would expect from unsigned subtraction.

The following C example demonstrates that what we have been discussing is indeed the case,

```
#include <stdio.h>

int main() {
    unsigned char x=0xEE, y=0x35, z;

    z = x - y;
    printf("%X\n", z);

    z = y - x;
    printf("%X\n", z);

    z = 0 - 1;
    printf("%X\n", z);
}
```

The output of this program is,

$$B9_{16} = 10111001_2$$
$$47_{16} = 01000111_2$$
$$FF_{16} = 11111111_2$$

which is exactly what we saw in the examples above.

Now that we know how to subtract in binary we can examine a useful trick involving AND. If one bit in a number is set this implies that the number is a power of two since every position in a binary number is, by definition, a power of two. If we know which bit we want to test for, which power of two, it is straightforward to use a mask and check that bit. But, what if we wanted to know if the number in question was any power of two? We can use AND here along with an observation about bits that are on in a number that is a power of two. For example, if the number we want to test is $00100000_2 = 32$ we see that it is a power of two and only one bit is on. Now, subtract one from this number. In this case, we will get $00011111_2 = 31$. What has happened is that the single bit that was on is now off and some of the bits that were off are now on. Finally, what happens if we AND these two values together? We get,

$$
\begin{array}{r}
00100000 = 32 \\
\text{AND} \quad 00011111 = 31 \\
\hline
00000000 = 00
\end{array}
$$

which is exactly zero. From this we see that we will only get exactly zero when one of two conditions are met: either the number is itself zero or it is a power of two which had only one bit set. This is nicely captured in a simple C function,

```
unsigned char is_power_of_two(unsigned char n) {
    return (n == 0) ? 0
                    : (n & (n-1)) == 0;
}
```

which returns 1 if the argument is a power of two and 0 otherwise. The function checks if the argument is zero, if so, return 0. If not, then check whether n & (n-1) is exactly 0. If it is, the expression is true and the function returns 1 to indicate a power of two, otherwise it returns 0. While written for unsigned char data type the function will work for any unsigned integer type.

We've looked in detail at addition and subtraction, now we turn our attention to multiplication and division. Modern microprocessors perform multiplication and division as operations in hardware. This is a very good thing but makes it difficult for us in a sense so we will, as before, look at more primitive approaches which might be used in small 8-bit microcontrollers that lack hardware instructions for multiplication and division. To keep things simple we will illustrate the algorithms in C even though this would never be done in practice.

Since multiplication is really repeated addition one approach to finding $n \times m$, where n and m are unsigned integers, would be to add n to itself m times or vice versa. Naturally, it would make sense to run the loop as few times as possible so we would loop over the smaller of n or m adding the other number. In C we have,

```
unsigned short mult1(unsigned char n, unsigned char m) {
    unsigned char i;
    unsigned short ans = 0;

    if (n < m) {
        for(i=0; i < n; i++)
            ans += m;
    } else {
        for(i=0; i < m; i++)
            ans += n;
    }

    return ans;
}
```

which leaves the product of the 8-bit numbers in n and m in the now possibly 16-bit value p. Why is the product possibly 16-bits? Because the largest possible number we can get by multiplying two 8-bit numbers requires 16-bits to store it in memory since $255 \times 255 = 65025$ which is above $2^8 - 1 = 255$ meaning it needs more than 8-bits to store but is below $2^{16} - 1 = 65535$ which is the maximum for a 16-bit unsigned integer.

Is this really a good way to multiply numbers, however? Probably not. The loop needs to be repeated for the smaller of n or m which may be up to 255 times. Given we must do 16-bit addition inside the loop, recalling the example above, we see that the simple multiplication may turn into many thousands of individual machine instructions. Surely we can do better than this? To answer this question, let's look a little more closely at multiplication in binary as we might do it by hand,

$$
\begin{array}{r}
00010100 \quad 14_{16} = \quad 20 \\
\times \quad \underline{00001110 \quad 0E_{16} = \quad 14} \\
00000000 \\
00010100 \\
00010100 \\
+ \quad \underline{00010100 } \\
\overline{00100011000 \quad 118_{16} = 280}
\end{array}
$$

in which we see that if the binary digit in the multiplier is 0 we simply copy down all zeros and if it is a 1 we copy the multiplicand lining it up beneath the multiplier bit as we would do in decimal. Then, again as in decimal multiplication, we add all the partial products to arrive at the final answer. For simplicity we did not write the leading zeros which would be present if showing all 16-bits of the result.

This method suggests a possible improvement over our existing multiplication function, mult1. Rather than repeatedly add the multiplier or multiplicand, we can copy the process just shown by shifting the multiplicand into position and adding it to the partial product if the multiplier bit is 1 otherwise ignore those that are 0. This leads to a second multiplication function in C,

```
unsigned short mult2(unsigned char n, unsigned char m) {
    unsigned char i;
    unsigned short ans = 0;

    for(i=0; i < 8; i++) {
        if (m & 1) {
            ans += n << i;
        }
        m >>= 1;
    }

    return ans;
}
```

which, when compared to mult1 and run ten million times proves to be about 1.6x faster. Let's look at what mult2 is actually doing.

We are multiplying two 8-bit numbers so we need to look at each bit in the multiplier, m. This is the source of the for loop. The if statement AND's the multiplier with 1 to extract the lowest bit. If this bit is set, we want to add the multiplicand, n, to the partial product stored in ans. Note, though, that before we add the multiplicand, we need to shift it up to the proper bit position. Note also that this works because the result of the operation is a 16-bit value which will not lose any of the bits of n when we do the shift. Regardless of whether we add anything to the partial product we need to shift the multiplier down one bit so that in the next pass through the loop the if will be looking at the next highest bit of the original m. Lastly, we see that there are no side-effects to this function because C passes all arguments by value meaning n and m are copies local to the function..

The speed improvement between mult1 and mult2 becomes much more dramatic when we move from multiplying two 8-bit numbers to multiplying two 16-bit numbers. To do this, we take the source for mult1 and mult2 and replace all instances of unsigned char by unsigned short and all instances of unsigned short by unsigned int. Lastly, we change the loop limit in mult2 from 8 to 16 since we are multiplying two 16-bit numbers. When this done we see that mult2 is nearly 3500x faster than mult1 for the same arguments (assuming both to be near the limit of 65535).

What about division? We cover two operations with division since the algorithm returns the quotient and any remainder. The operations are integer division (/) which returns the quotient and modulo (%) which returns the remainder. For example, we need an algorithm that produces these answers,

$$123 \; / \; 4 = 30$$
$$123 \; \% \; 4 = \;\; 3$$

since $123/4 = 30$ with a remainder of 3.

We could implement division by repeated subtraction. If we count the number of times we can subtract the divisor from the dividend before we get a partial result that is less than the divisor we will have the quotient and the remainder. We might code this in C as,

```
unsigned char div1(unsigned char n,
                   unsigned char m,
                   unsigned char *r) {
    unsigned char q=0;

    *r = n;

    while (*r > m) {
        q++;
        *r -= m;
    }

    return q;
}
```

and test it with,

```
int main() {
    unsigned char n=123, m=7;
    unsigned char q,r;

    q = div1(n, m, &r);
    printf("quotient=%d, remainder=%d\n", q,r);
}
```

which prints quotient=30, remainder=3 which is the answer we are expecting.

This function requires three arguments since we want to return the quotient as the function value and the remainder as a side-effect value. This is why we pass the remainder as a third argument using a pointer. Inside of div1 we set the remainder (r) to our dividend and continually subtract the divisor (m) until we get a result less than the divisor. While doing this we keep count of the number of times we subtract in q which we return as the quotient.

Like our `mult1` example above, `div1` is an inefficient way to implement division. What happens if the dividend is large and the divisor is small? We must loop many times in that case before we are done. The problem is even worse if we use integers larger than 8-bits. What to do, then?

Just as we did for multiplication, let's look at binary division by hand. Unlike decimal long division, binary division is rather simple, either the divisor is less than or equal to the dividend in which case the quotient bit is 1, otherwise, the quotient bit is 0; there are no trial multiplications. Dividing $123 = 01111011_2$ by $4 = 100_2$ in this way gives,

$$
\begin{array}{r}
00011110 \\
100\,\overline{)01111011} \\
0 \\
\overline{01} \\
0 \\
\overline{011} \\
0 \\
\overline{0111} \\
100 \\
\overline{0111} \\
100 \\
\overline{0110} \\
100 \\
\overline{0101} \\
100 \\
\overline{11} \\
0 \\
\overline{11}
\end{array}
$$

with $00011110_2 = 30$ and a remainder of $11_2 = 3$ as expected.

As we have seen several times already, modern microprocessors implement such a fundamental operation as division in hardware, but we can look at unsigned division in the way it might be implemented in a more primitive microprocessor or microcontroller. For simplicity, we again implement the algorithm in C. With all of this in mind, we get Fig. 2.5 which will require a bit of explanation.

The key to understanding what Fig. 2.5 is doing it to observe that binary division by hand is really a matter of testing whether or not we can subtract the divisor from the partial dividend. If so, we set a one in that bit of the quotient, otherwise we set a zero. The algorithm of Fig. 2.5 is setup for 8-bit division using 8-bit dividends and divisors, and by implication quotients. Therefore, we need to examine all eight bits of the dividend starting with the highest bit. To do this, and to save space as this algorithm is often implemented in hardware, we take advantage of the fact that C passes arguments by value and use n to be both dividend and quotient. When we have examined all eight bits the value in n will be the quotient. We can do this because as we look at each bit of the dividend we shift it out to the left while shifting in the new bit of the quotient from the right.

```
 1| unsigned char div2 (unsigned char n,
 2|                     unsigned char m,
 3|                     unsigned char *r) {
 4|     unsigned char i;
 5|
 6|     *r = 0;
 7|
 8|     for(i=0; i<8; i++) {
 9|         *r = (*r << 1) + ((n & 0x80) != 0);
10|         n <<= 1;
11|
12|         if ((*r-m) >= 0) {
13|             n |= 1;
14|             *r -= m;
15|         }
16|     }
17|
18|     return n;
19| }
```

Fig. 2.5 Shift, test, and restore unsigned integer division

We store the remainder in r and pass it back out of the function by using a pointer. To start the division process we set r to zero and the quotient to the dividend in n. Since n already has the dividend there is no explicit operation to do this, we get it for free. If we call div2(123, 4, &r) to continue with our running division example, the state of affairs in binary after the execution of line 6 in Fig. 2.5 is,

i	r	n	m
undefined	00000000	01111011	100

where the dividend is in n and the remainder is zero. We next hit the loop starting in line 8. This loop executes eight times, once for each bit of the dividend. Lines 9 and 10 perform a double left shift. This is the equivalent of treating r and n as a single 16-bit variable with r the high order bits. First we shift r one bit to the left (*r << 1) and then comes the rather cryptic expression,

$$((n \ \& \ 0x80) \ != \ 0)$$

which tests whether the highest bit in n, our dividend and quotient is set. Recall our discussion of AND above. If it is set, we add it into r. This is because we are about to left shift n one bit and if the bit we are shifting out of n is set, we need to move it to r to complete the virtual 16-bit shift of r and n. We then shift n in line 10.

Line 12 checks whether or not we can subtract the divisor in m from the partial dividend which is being built, bit by bit, in r. If we can, we set the first bit of n, our quotient, in line 13 and then update the partial dividend by subtracting the divisor in line 14. Recall, we are examining the dividend one bit at a time by moving it into r. We are simultaneously storing the quotient in n one bit at a time by putting it in on the right side. Since we already shifted to the left the first bit in n is always zero, we only update it if the subtraction succeeds. After this first pass through the loop we have,

i	r	n	m
0	00000000	11110110	100

with the first bit of the quotient, a zero, in the first bit of n. If we continue through the loop we will get the following sequence of values,

i	r	n	m
1	00000001	11101100	100
2	00000011	11011000	100
3	00000011	10110001	100
4	00000011	01100011	100
5	00000010	11000111	100
6	00000001	10001111	100
7	00000011	00011110	100

where we end with a quotient of 30 in n which is the return value of the function and a remainder in r of 3. Notice how n changes as we move through the loop. Each binary digit is shifted into r from the right as the new quotient bits are assigned from the left until all bits are examined. This algorithm, unlike the div1 example, operates in constant time. There are a fixed number of operations needed regardless of the input values.

2.3.6 Square Roots

We briefly consider here a simple integer square root algorithm which makes use of an interesting mathematical fact. This algorithm works by counting the number of times an ever increasing odd number can be subtracted before reaching or going below zero. The algorithm itself is easy to implement in C,

```
unsigned char sqr(unsigned char n) {
    unsigned char c=0, p=1;

    while (n >= p) {
        n -= p;
        p += 2;
        c++;
    }

    return c;
}
```

where we again make use of the fact that C passes arguments by value which allows us to modify n in the function without changing it outside of the function. Our count, which will be the square root of n, is initialized to zero in c. We start our odd number in p at 1 and then move to 3, 5, 7, and so on. The while loop is checking to see if our n value is still larger or the same as p and if so, we subtract p and count one more subtraction in c. When n is less than p we are done counting and return c as the square root. Naturally, this algorithm is only approximate by underestimating when n is not actually a perfect square.

If we call `sqr` with `64` as the argument, we get the following sequence of values in the `while` loop,

n	p	c
63	3	1
60	5	2
55	7	3
48	9	4
39	11	5
28	13	6
15	15	7
0	17	8

where the final value of n is zero since `64` is a perfect square and c is `8`, which is the square root of `64`. We see that the algorithm works, but why?

The trick is the observation that the sum of the sequence of odd numbers is always a perfect square. For example,

$$1 = 1$$
$$1 + 3 = 4$$
$$1 + 3 + 5 = 9$$
$$1 + 3 + 5 + 7 = 16$$
$$1 + 3 + 5 + 7 + 9 = 25$$

or more compactly,

$$\sum_{i=1}^{n} 2i - 1 = n^2$$

where n^2 is the argument to `sqr` and n is the square root.

2.4 What About Negative Integers?

In the previous section we took a thorough look at unsigned integers and the sorts of operations computers typically perform on them. Without a doubt, unsigned integers are the mainstay of computers, but often it is necessary to represent quantities that are less than zero. What do we do about that? In this section we examine three options, two in detail, for tracking the sign of an integer and performing operations with signed integers. We will naturally build on what we have learned about unsigned integers and bear in mind that, as before, while we may show examples using 8-bit numbers for simplicity, everything immediately translates to numbers with more bits, be they 16, 32, or larger.

The most common techniques for handling signs are sign-magnitude, one's complement, and two's complement.

2.4.1 Sign-Magnitude

Perhaps the most natural way to represent the sign of an integer is to reserve one bit of its representation for the sign and this is precisely what early computers did. If we decide that we will keep the highest-order bit of the number for the sign we can use the remainder of the bits to represent the magnitude as an unsigned integer and this is the *sign-magnitude* form,

```
01111111 =  127
...
00000010 =    2
00000001 =    1
00000000 =    0
10000001 =   -1
10000010 =   -2
...
11111111 = -127
```

This seems to be a perfectly reasonable way to store a signed integer but notice a few things,

- We lose range in terms of magnitude. An unsigned 8-bit number can take values from 0 to 255 while a sign-magnitude number is restricted to -127 to +127. As we will see, keeping track of the sign always results in a loss of magnitude range.
- There are two ways to represent zero: +0 = 00000000 and -0 = 10000000. This seems unnecessary and wasteful of a bit pattern.
- Arithmetic becomes more tedious since we need to bear the sign of the number in mind at all times. It would be nice to be able to do some arithmetic without requiring separate logic for the sign.

For the reasons given above, especially the additional hardware logic, the sign-magnitude representation for integers has been abandoned by modern computer systems. Let us now turn our attention to possible replacements.

2.4.2 One's Complement

Our first candidate for a suitable replacement to the sign-magnitude form is called *one's complement*. In this notation we represent negative numbers by taking the positive form and calculating the one's complement. The one's complement is simple to do, just negate (logical NOT) every bit in the positive form of the number. So, we have,

```
01111111 =  127
...
00000010 =    2
00000001 =    1
00000000 =    0
11111110 =   -1
11111101 =   -2
...
10000000 = -127
```

which again seems good in that we can look at the highest order bit to see if the number is negative or not and as we will shortly see we can use this notation for arithmetic without too much trouble, but we do still have two ways to represent zero since $00000000 \to 11111111$.

2.4.3 Two's Complement

The one's complement of a positive number is the bit pattern we get when we change all the zero bits to one and all the one bits to zero. The *two's complement* of a positive number is the bit pattern we get when we take the one's complement and then add one to it. This is the notation that has been accepted as the way to represent negative integers and the advantages will be come clear when we look at operations on signed integers. As with one's complement integers, a positive two's complement integer is represented in just the same way as an unsigned integer. With two's complement we have,

```
01111111 =  127
...
00000010 =    2
00000001 =    1
00000000 =    0
11111111 =   -1
11111110 =   -2
...
10000000 = -128
```

where we now have only one way to represent zero since,

$$00000000 \to \quad 11111111 \quad \to \quad 00000000$$
$$\textit{positive} \quad \textit{one's complement} \quad \textit{two's complement}$$

since adding one to 11111111 maps back around to 00000000 with the overflow bit ignored. Additionally, we have increased our range by one since we can represent numbers from -128 to $+127$ instead of -127 to 127 as with one's complement or sign-magnitude.

2.5 Operations on Signed Integers

We would like to be able to perform operations on signed integers. The bit level operations like AND, OR and NOT work the same way with signed integers as with unsigned integers. To these operators, the bits are just bits, the "fact" of a negative integer is just a convention forced on certain bit patterns. Since this is the case, we need only look at how to compare negative integers, how to perform arithmetic on negative integers, and, as a special operation, how to deal with the sign of a two's complement integer when changing the number of bits used to represent the number. Let us first start with comparing two signed integers.

2.5.1 Comparison

Comparison of two signed integers, A and B, implies determining which relational operator, $<$, $>$, or $=$, should be put between them. When we compared unsigned integers we looked at the bits from highest to lowest. We still to that for signed integers but we need to first consider the signs of the two numbers. If the signs differ we know very quickly the relationship between the numbers without considering all the bits. If the signs match, either both positive or both negative, we need to look at the magnitude bits to see where they might be different. This will tell us the relationship between the two numbers. Naturally, if the bits are all the same, position for position, then the two numbers are equal and use the exact same bit pattern.

We can use a C function like the one in Fig. 2.6, with included helper function to determine whether or not a particular bit position is on, to compare two signed numbers in two's complement notation. Note that we are now working with variables of type signed char which are 8-bit signed integers.

The helper function (bset, lines 1–4) returns 1 if the n-th bit of v is on, otherwise it returns a 0. It uses the shift and AND mask trick we saw above to test a bit position value. The main function, scomp, looks first at the signs of the arguments (lines 10–13) and returns the relationship if they differ. If the sign bit of a is zero, a is positive. If the sign bit of b is one, then b is negative and a must be greater than b so return 1 to indicate a > b. If the signs are reversed, a is less than b so return -1 to indicate a < b.

If the signs of a and b match, either positive or negative, we then look at the remaining bits from highest to lowest to see where there are any differences. This is the loop of lines 15 through 20 in Fig. 2.6. If we find a bit position where a has a zero and b has a one we know that a < b must be true so we return -1. Likewise, if we find that a is one and b is zero at that bit position we know that a > b so we return 1. Finally, if we make it through all the bits and find no differences the two integers are equal so we return 0. With a comparison function like scomp it is easy to create predicate functions checking for equal, less than and greater than. Recall that a predicate function is a function that returns true or false. For example,

```
 1 signed char bset(signed char v,
 2                  signed char n) {
 3     return (v & (1 << n)) != 0;
 4 }
 5
 6 signed char scomp(signed char a,
 7                   signed char b) {
 8     unsigned char i;
 9
10     if ((bset(a,7) == 0) && (bset(b,7) == 1))
11         return 1;
12     if ((bset(a,7) == 1) && (bset(b,7) == 0))
13         return -1;
14
15     for(i=0; i<7; i++) {
16         if ((bset(a,6-i) == 0) && (bset(b,6-i) == 1))
17             return -1;
18         if ((bset(a,6-i) == 1) && (bset(b,6-i) == 0))
19             return 1;
20     }
21
22     return 0;
23 }
```

Fig. 2.6 Comparison of two signed integers a and b. The function returns 0 if a == b, 1 if a > b and -1 if a < b

```
unsigned char isEqual(signed char a,
                      signed char b) {
    return scomp(a,b) == 0;
}

unsigned char isLessThan(signed char a,
                         signed char b) {
    return scomp(a,b) == -1;
}

unsigned char isGreaterThan(signed char a,
                            signed char b) {
    return scomp(a,b) == 1;
}
```

define functions which return true (1) when $a = b$, $a < b$, and $a > b$, respectively.

2.5.2 Arithmetic

Let's take a look at the basic arithmetic operations $(+, -, \times, \div, \text{ and } \%)$ as they apply to signed numbers. For addition and subtraction we will consider both one's and two's complement negative numbers to show why two's complement is often preferable. We focus on operations involving negative numbers as operations involving positive numbers follow the techniques described earlier in the chapter for unsigned integers.

Addition and Subtraction. Addition in one's complement notation is nearly identical to unsigned addition with one extra operation should there be a final carry. To see this, let's take a look at adding two negative integers represented in one's complement,

$$
\begin{array}{r}
11000000 \quad -63 \\
+ \quad 11000010 \quad -61 \\
\hline
1\ 10000010 \quad -125 \\
10000011 \quad -124
\end{array}
$$

(one's complement)

where the carry at the end, shown by the extra 1 on the left, is added back into the result to get the correct answer of -124. This adding in of any carry is called the *end-around carry* and is the extra twist necessary when adding one's complement numbers.

The same addition in two's complement notation produces a carry which we ignore,

$$
\begin{array}{r}
11000001 \quad -63 \\
+ \quad 11000011 \quad -61 \\
\hline
1\ 10000100 \quad -124
\end{array}
$$

(two's complement)

since we see that 10000100 is -124 by making it positive,

$$
\begin{array}{r}
01111011 \\
+\ 00000001 \\
\hline
01111100 \quad 124
\end{array}
$$

Addition of a positive and negative number works in the same way for both one's and two's complement numbers. For example, in one's complement we have,

$$
\begin{array}{r}
+ \quad 01111100 \quad 124 \\
11000010 \quad -61 \\
\hline
1\ 00111110 \quad 62 \\
00111111 \quad 63
\end{array}
$$

(one's complement)

where we have again made use of the end-around carry to give us the correct answer. The two's complement version is similar,

$$
\begin{array}{r}
+ \quad 01111100 \quad 124 \\
11000011 \quad -61 \\
\hline
1\ 00111111 \quad 63
\end{array}
$$

(two's complement)

where we again ignore the carry and keep only the lower eight bits. Recall, we are giving all examples as signed or unsigned 8-bit numbers. If we were working with 16-bit or 32-bit numbers we would keep that many bits in the answer.

Computers implement subtraction by negation and addition. This allows for only one set of hardware circuits to be used for both operations. With that in mind, subtraction becomes particularly simple. If we want to calculate $124 - 61 = 63$ we actually calculate $124 + (-61) = 63$ which is exactly the example calculated above. For calculation by hand it is helpful to think of subtraction as an actual operation but, as we see here, when done with the appropriate notation for negative numbers, subtraction is really an "illusion" and is nothing more than addition with a negative number.

While addition and subtraction are straightforward, especially with two's complement notation, we have to consider one question: what happens if the result of the operation does not fit in the number of bits we are working with? For our examples, this means that the result does not fit in eight bits taking the sign into account. Let's consider only two's complement numbers. We already saw in the examples above that we could ignore the carry to the 9th bit and saw that the lower eight bits were correct. Is this always the case?

A signed 8-bit two's complement number has a range from -128 to 127. If we attempt an operation that falls outside of this range we will not be able to properly represent the answer. We call this condition an *overflow*. How can we detect this? By following two simple rules,

1. If the sum of two positive numbers is negative, overflow has happened.
2. If the sum of two negative numbers is positive, overflow has happened.

We need not worry about the sum of a positive and negative number because both the positive and negative number are already in the allowed range and it is impossible, because of the difference in sign, for the sum to be outside the allowed range. This is why we ignored the last carry bit when adding -61 to 124. Let's look at cases that prove the rules. If we try to calculate $124 + 124 = 248$ we know we will have trouble because 248 is greater than 127 which is the largest 8-bit positive two's complement number. We get,

```
+ 01111100   124
  01111100   124      (two's complement)
  _____
  11111000    -8
```

which is clearly a wrong answer. According to our rule for addition of two positive numbers we know overflow has happened because the sign bit, bit 7, is one, indicating a negative answer. Similarly, two large negative numbers added will prove our second rule,

```
+ 10000100   -124
  10000100   -124      (two's complement)
  _____
  00001000      8
```

where we have ignored the carry to the 9th bit. We see that the result is positive since bit 7 is zero. This proves our second rule and we know that overflow has happened.

Multiplication. We now consider multiplication of signed integers. One approach
to signed multiplication would be to make use of the rules for products to track the
sign of the result. If we do this, we can make any negative number positive, do unsigned integer multiplication as described above in Sect. 2.3.5, and negate the result
if necessary to make it negative. This approach will work for both one's and two's
complement numbers. As we recall from school, when multiplying two numbers
there are four possible scenarios related to the signs,

1. *positive × positive = positive*
2. *positive × negative = negative*
3. *negative × positive = negative*
4. *negative × negative = positive*

with this in mind it is simple to extend our `mult2` example from Sect. 2.3.5 to
check the signs of the inputs and negate the negative numbers to make them positive
before multiplying. Then, the result can be negated to make it negative if the result
should be negative. In C this gives us Fig. 2.7

```
 1| signed short signed_mult2(signed char n, signed char m) {
 2|     unsigned char i, s=0;
 3|     signed short ans=0;
 4|
 5|     if ((n > 0) && (m < 0)) {
 6|         s = 1;
 7|         m = -m;
 8|     }
 9|     if ((n < 0) && (m > 0)) {
10|         s = 1;
11|         n = -n;
12|     }
13|     if ((n < 0) && (m < 0)) {
14|         n = -n;
15|         m = -m;
16|     }
17|
18|     for(i=0; i < 8; i++) {
19|         if (m & 1) ans += n << i;
20|         m >>= 1;
21|     }
22|
23|     if (s) ans = -ans;
24|     return ans;
25| }
```

Fig. 2.7 Unsigned integer multiplication modified for signed numbers

where the main loop in lines 18 through 21 has not changed but before we run it we
check the signs of the inputs to see if we need to negate any negative numbers to
make them positive. The variable s holds the flag to tell us that the answer needs to

be negative. We initially set it to 0 with the assumption that the inputs, n and m, will both be positive. In lines 5 through 16 we check this assumption. If n is positive and m is negative, we set s in line 6 and make m positive. Likewise, in line 9 we check to see if n is negative and m is positive and make n positive if this is the case. We also set the flag in s since we know the answer needs to be negative. Lastly, if both n and m are negative we make them both positive and leave the flag in s unset. Then we multiply as before. In line 23 we check to see if the negative flag is set, if so, we make the answer negative before returning it.

In this example we have taken advantage of the fact that the C compiler will properly negate a value as in line 7 regardless of the underlying notation used for negative numbers. We know, however, in practice that this will typically be two's complement. Can we multiply numbers directly in two's complement? Yes, in fact, there are several existing algorithms which to exactly that. Let's consider one of the more popular of them, the Booth algorithm [1] which was developed by Andrew Booth in 1950. A C implementation of this algorithm for multiplying two signed 8-bit integers is given in Fig. 2.8. Let's take a look at what it is doing.

```
 1  signed short multb8(signed char m, signed char r) {
 2      signed int A, S, P;
 3      unsigned char i;
 4
 5      A = m << 9;
 6      S = (-m) << 9;
 7      P = (r & 0xff) << 1;
 8
 9      for(i=0; i < 8; i++) {
10          switch (P & 3) {
11              case 1:    // 01
12                  P += A;
13                  break;
14              case 2:    // 10
15                  P += S;
16                  break;
17              default:   // 11 or 00
18                  break;
19          }
20          P >>= 1;
21      }
22
23      return P>>1;
24  }
```

Fig. 2.8 The Booth algorithm for multiplication of two's complement 8-bit integers

Booth's essential insight was that when we multiply two binary numbers a string of ones can be replaced by a positive and negative sum in the same way that $16 \times 6 = 16 \times (8 - 2)$ but since we are in binary we can always write any string of ones as the next higher bit minus one. So, we have,

$$00011100 = 00100000 + 000000\text{-}10$$

where we have written a -1 for a specific bit position to indicate subtraction. This means, if we scan across the multiplicand and see that at bit position i and $i-1$ there is a 0 and 1, respectively, we can add the multiplier. Similarly, when we see a 1 and 0, respectively, we can subtract the multiplier. At other times, we neither add nor subtract the multiplier. After each pair of bits, we shift to the right.

In Fig. 2.8 we initialize the algorithm by setting A to the multiplier, m, shifted nine places to the left, and similarly set S to the negative of the multiplier (two's complement form), also shifted nine positions to the left. The product, P, is initialized to the multiplicand in r but shifted one position to the left. This is done in lines 5 through 7. Why all the shifting? We are multiplying two eight bit signed numbers so the result may have as many as 16 bits, hence using signed int for A, S and P. This is the origin of eight of the nine bit positions. The extra bit position for A and S, and the single extra bit for P (line 7), is so that we can look at the last bit position and the one that would come after which we always set to zero. This means that the last bit position, bit 0, and the next, bit -1, could be 1 and 0 to signal the end of a string of ones. We mask the multiplicand, r, with 0xFF to ensure that the sign is not extended when we move from the signed char to signed int data type. See the next section for a description of sign extension.

The loop in lines 9 through 21 examines the first two bits of P, which are the two we are currently considering, and decides what to do based on their values. Our options are,

bit i	bit $i-1$	operation
0	0	*do nothing*
0	1	add multiplier to product
1	0	subtract multiplier from product
1	1	*do nothing*

which is reflected in the switch statement of line 10. The phrase P & 3 masks off the first two bits of P, which is what we want to examine. After the operation, we shift the product (P) to the right to examine the next pair of bits. When the loop finishes, we shift P once more to the right to remove the extra bit we added at the beginning in line 7. This completes the algorithm and we have the product in P, which we return. This algorithm substitutes a starting add and ending subtraction for what might be a long string of additions for each 1 bit in a string of 1 bits. Also, when not adding or subtracting, we simply shift bits. This makes the algorithm particularly efficient.

Sign Extension and Signed Division. Just as we did for multiplication above, we can modify the unsigned integer algorithm for division in Fig. 2.5 to work with signed integers by determining the proper sign of the output, then making all arguments positive and dividing, negating the answer if necessary. However, before we do that, let's take a quick look at sign extension.

Sign Extension. What happens if we take an 8-bit number and make it a 16-bit number? If the number is positive, we simply set the upper eight bits of the new 16-bit number to zero and the lower eight bits to our original number like so,

$$00010110 \rightarrow 0000000000010110$$

which is pretty straightforward. Now, if we have a negative 8-bit signed integer in two's complement notation we know that the leading bit will be a 1. If we simply add leading zeros we will get,

$$11011101 \rightarrow 0000000011011101$$

which is no longer a negative number because the leading bit is now a 0. To avoid this problem and preserve the value we extend the sign when we form the 16-bit number by making all the new higher bits 1 instead of 0,

$$11011101 \rightarrow 1111111111011101$$

which we know is the same value numerically and we can check it by looking at the magnitude of the number. Recall, we convert between positive and negative two's complement by flipping the bits and adding one. So, the 8-bit version becomes,

$$11011101 \rightarrow 00100010 + 1 \rightarrow 00100011 = 35_{10}$$

and the 16-bit version becomes,

$$1111111111011101 \rightarrow 0000000000100010 + 1 \rightarrow 0000000000100011 = 35_{10}$$

which means that both bit patterns represent -35 as desired. We intentionally frustrated sign extension in Fig. 2.8 by masking r with 0xFF before assigning it to P which was a 32-bit integer.

Signed Division. Figure 2.5 implements unsigned division. If we track the signs properly we can modify it to work with signed integers. Division actually returns two results. The first is the quotient and the second is any remainder. The sign we should use for the quotient is straightforward enough,

Dividend	Divisor	Quotient
+	+	+
+	−	−
−	+	−
−	−	+

ambiguity arises when we think about what sign to apply to the remainder. It turns out that different programming languages have adopted different conventions. For example, C chooses to make the remainder have the same sign as the dividend while Python gives the remainder the sign of the divisor. Unfortunately, the situation is more complicated still. When dividing negative numbers we are often returning an approximate quotient (unless the remainder is zero) and that approximate quotient has to be rounded in a particular direction. All division algorithms in programming languages satisfy $d = nq + r$ which means that the quotient, q, times the divisor, n, plus the remainder, r, equals the dividend, d. However, we have choices regarding how to set the signs and values of q and r. There are three options,

1. *Round towards zero.* In this case, we select q to be the integer closest to zero that when multiplied by n is less than or equal to d. In this case, if d is negative, r will also be negative. For example, $-33/5 = -6$ with a remainder of -3 so that $-33 = 5(-6) + (-3)$. This is the option used by C.
2. *Round towards negative infinity.* Here we round the quotient down and end up with a remainder that has the same sign as the divisor. In this case, $-33/5 = -7$ with a remainder of 2 giving $-33 = 5(-7) + 2$. This is the option used by Python.
3. *Euclidean definition.* This definition makes the remainder always positive. If $n > 0$ then $q = floor(d/n)$ and if $n < 0$ then $q = ceil(d/n)$. In either case, r is always positive, $0 \leq r < |n|$.

Let's make these definitions more concrete. The table below shows the quotient and remainder for several examples in both C and Python. These will give us an intuitive feel for how these operations work.

Dividend	Divisor	Quotient	Remainder	
33	7	4	5	C
		4	5	Python
-33	7	-4	-5	C
		-5	2	Python
33	-7	-4	5	C
		-5	-2	Python
-33	-7	4	-5	C
		4	-5	Python

We see that differences only arise when the signs of the dividend and divisor are opposite. It is here that the C choice of rounding towards zero and the Python choice of rounding towards negative infinity come into play. The C choice seems more consistent at first because it always results in quotients and remainders with the same magnitude, only the signs change, but from a mathematical point of view it is less desirable because certain expected operations do not give valid results. For example, to test whether or not an integer is even it is common to check if the remainder is zero or one when dividing by two. If the remainder is zero, the number is even, if one, it is odd. This works in Python for negative integers since `-43 % 2 = 1` as expected, but in C this fails since we get `-43 % 2 = -1` because of the convention to give the remainder the sign of the dividend.

With all of the above in mind, we can now update our `div2` algorithm in Fig. 2.5 to handle signed integers. We show the updated algorithm, now called `signed_div2`, in Fig. 2.9. Let's look at what has changed.

The actual division algorithm in lines 20 through 30 is the same as in Fig. 2.5 since we are still performing unsigned division. Lines 6 through 18 check on which of the arguments, the dividend (n) or divisor (m) or both, are negative. We use two auxiliary variables, `sign` and `rsign`, to track how we deal with the sign of the answer. If the dividend is negative but the divisor is positive the quotient should be negative so we set `sign` to 1. If the dividend is positive but the divisor is negative, we also need to make the quotient negative. If both the dividend and divisor are

```
 1  signed char signed_div2(signed char n,
 2                          signed char m,
 3                          signed char *r) {
 4      unsigned char i, sign=0, rsign=0;
 5
 6      if ((n < 0) && (m > 0)) {
 7          sign = rsign = 1;
 8          n = -n;
 9      }
10      if ((n > 0) && (m < 0)) {
11          sign = 1;
12          m = -m;
13      }
14      if ((n < 0) && (m < 0)) {
15          rsign = 1;
16          n = -n;
17          m = -m;
18      }
19
20      *r = 0;
21
22      for(i=0; i<8; i++) {
23          *r = (*r << 1) + ((n & 0x80) != 0);
24          n <<= 1;
25
26          if ((*r-m) >= 0) {
27              n |= 1;
28              *r -= m;
29          }
30      }
31
32      if (sign) n = -n;
33      if (rsign) *r = -*r;
34
35      return n;
36  }
```

Fig. 2.9 Shift, test, and restore unsigned integer division updated to handle signed integers

negative, the quotient is positive. In all cases we make the dividend and divisor positive after we know how to account for the sign of the quotient. For the remainder, we use the variable rsign to decide when to make it negative. The division algorithm itself will make the remainder, in *r, positive but in order for our answer to be sound we must sometimes make *r negative. When to do this? A sound answer will always satisfy,

$$n = m \times q + r$$

so if the dividend in n was negative, we will require the remainder to be negative as well. In this case we follow the C convention.

If we run `signed_div2` on n = 123 and m = 4 with all combinations of signs, we get the following output,

n	m	q	r	$m \times q + r$
123	4	30	3	4(30)+3
-123	4	-30	-3	4(-30)-3
123	-4	-30	3	-4(-30)+3
-123	-4	30	-3	-4(30)-3

where the column on the right shows that our choice of sign for the remainder is correct.

In this section we have reviewed implementations of signed arithmetic on integers. In some cases we were able to build directly on existing algorithms for unsigned arithmetic while in some we worked directly in two's complement format. There is no need to talk about signed integer square root since the square root of a negative number is imaginary and we are not yet ready to work with complex numbers.

2.6 Binary-Coded Decimal

Binary-Coded Decimal (BCD) numbers make use of specific bit patterns corresponding to the digits $0 \ldots 9$ in order to store one or two decimal digits in each byte. Storing numbers in this format allows for decimal operations in place of binary and, indeed, some early microprocessors such as the Western Digital 6502 had BCD modes. In this section we describe how to encode numbers in BCD and how to do simple arithmetic with those numbers.

2.6.1 Introduction

As we saw earlier in this chapter, working with numbers expressed in binary can be cumbersome. Binary is the natural base for a computer to use given its construction but humans, with ten fingers and ten toes, generally prefer to work in decimal. One possibility is to encode decimal digits in binary and let the computer work with data in this format. Binary-coded decimal does just this. For our purposes we will work with what is generally called *packed BCD* where we use each nibble of a byte to represent exactly one decimal digit. Historically, other sizes were used, one digit per byte (unpacked) for example, or even other values, typically with early computers. In addition, while any set of distinct bit patterns can be used to represent decimal digits we will use the obvious choice, $0 = 0000, 1 = 0001, \ldots, 9 = 1001$, so there is a direct conversion between a decimal digit and its BCD bit pattern. The remaining six bit patterns, $1010 \ldots 1111$ are not used or allowed in properly formatted BCD numbers. In the previous sections we ignored the difficulties in converting human entered decimal data to binary and vice versa. BCD simplifies this process. BCD is no longer frequently used but it will turn up again when we consider some of the more specialized ways in which computers represent numbers.

If we consider only positive numbers, a single byte can represent decimal numbers from zero through 99 as so,

BCD	Decimal
0 0	0
0 1	1
0 2	2
1 0	10
1 1	11
...	
9 8	98
9 9	99

where the two digits in the *BCD* column on the left represent the two 4-bit nibbles of the byte. This works but does not allow for negative numbers. Historically, several variations were used for storing a sign with a BCD number. The form we will use is directly analogous to signed binary integers but instead of two's complement we will use *ten's complement* with a leading nibble to represent the sign. If the leading nibble is 0000 the number is positive, if 1001 the number is negative and the value is in *ten's complement format*. We use the leading sign nibble because unlike two's complement where the leading bit is always 1 if the number is negative, ten's complement has no such quick test.

In two's complement we negate a number by flipping all the digits making $0 \rightarrow 1$ and $1 \rightarrow 0$ and then add one. For the ten's complement of a decimal number we first find the nine's complement which involves subtracting the number from 99, for a single byte, and then add one. So, we see that the BCD number 51 can be thought of as -49 because,

$$99 - 51 = 48, \ 48 + 1 = 49$$

A multidigit BCD number is represented with multiple bytes. Endian issues arise again in this case and we chose to use big-endian so that it is easier to see the decimal numbers in the binary bit patterns. With this convention, including the sign nibble, the decimal number 123 is represented in BCD as,

$$\begin{array}{c|c|cc} + & 0 & 1 & 2 & 3 \\ 0000 & 0000 & 0001 & 0010 & 0011 \end{array}$$

where we need the leading 0 digit to fill out the byte representing the leading two digits. Similarly, using the negative sign nibble we can represent -732 as,

$$\begin{array}{c|c|cc} - & 0 & 2 & 6 & 8 \\ 1001 & 0000 & 0010 & 0110 & 1000 \end{array}$$

because $999 - 732 = 267, \ 267 + 1 = 268$ is the ten's complement of 732.

2.6.2 Arithmetic with BCD

Let's look at arithmetic using BCD numbers. We only consider addition and subtraction as multiplication and division are rarely performed in BCD. For performance reasons it is faster to convert from BCD to binary, do the multiplication or division in binary, and then convert the answer back to BCD.

Addition of two BCD numbers is straightforward at first glance. We simply add nibble by nibble since each nibble is a digit,

```
  123      0000 | 0001 0010 0011
+ 732    + 0000 | 0111 0011 0010
  855      0000 | 1000 0101 0101
```

where for each nibble we simply add in binary as if the numbers were unsigned binary integers. For this example, everything works out nicely. But consider this,

```
  123      0000 | 0001 0010 0011
+ 738    + 0000 | 0111 0011 1000
  861      0000 | 1000 0101 1011
```

here we have a small problem, the sum of the first digits gives us `1011` which is 11, not an allowed bit pattern in BCD. We see the source of the problem which is that $3 + 8 = 11$ so we have produced a carry. A simple way to deal with any carries is to take the resulting bit pattern, `1011`, and add six, `0110`, which will give us the correct bit pattern for the decimal digit and a carry of `1` for the next digit. This correction gives,

$$1011 + 0110 = 1\ 0001$$

with the carry separated from the remaining digit which is now the BCD representation of `1`. Applying this correction to the addition leads to the correct result,

```
  123      0000 | 0001 0010 0011
+ 738    + 0000 | 0111 0011 1000
           0000 | 1000 0101 1011
  861      0000 | 1000 0110 0001
```

where we have added the carry into the next digit to change the 5 to a 6. Why add six? There are 16 possible 4-bit nibbles but we are only using the first 10 of them for decimal digits. If we exceed 10 for any single digit adding six will wrap the bit pattern around to the digit we would get if we, in decimal, subtracted ten. It also leaves the carry set for the next digit. If adding the carry to the next higher digit results in 10 or greater the "add six" trick can be used again and repeated as many times as necessary until the BCD number is in the proper form, each time moving to the next higher digit position.

We have chosen to represent negative BCD numbers using ten's complement. This format works in the same way as two's complement so that in order to subtract we simply add. For example, consider,

```
 123      0000 | 0001 0010 0011
- 732    + 1001 | 0010 0110 1000
─────────────────────────────────
         1001 | 0011 1000 1011
-609     1001 | 0011 1001 0001
```

where the overflow in the first digit was addressed by adding six and then adding the carry to the next higher digit. The sign is negative and the BCD number reads as 391. Since the number is negative, we expect that 391 is the ten's complement form of -609 which is the actual answer. To see that it is, we negate it,

$$999 - 391 = 608, \quad 608 + 1 = 609$$

so we know that we have, in fact, reached the correct result.

2.6.3 Conversion Routines

Binary to BCD and BCD to binary conversion is straightforward. First, let's consider binary to BCD. This routine could be used during multiplication or division of BCD numbers or might be used on its own to prepare for output of a binary value as a decimal number. Once the value is in BCD conversion to ASCII for output is straightforward, simply examine each nibble, add it to 48 which is the ASCII code for "0" and output the resulting byte as an ASCII character.

Our binary to BCD conversion routine will convert an 8-bit unsigned binary number to a three digit BCD number. It goes by the amusing name of the "double dabble" algorithm and we present our implementation in Fig. 2.10. The routine itself is referred to briefly in [7].

What is this code doing? We have an unsigned 8-bit binary number passed in as b. We will return the three digit BCD number in p. Since a single byte can hold any value from zero to 255 we need exactly 12 bits to store the equivalent BCD number. Using an unsigned short as a return value gives us 16 bits, the top four of which we leave as zero. Inside the algorithm we work with p as an unsigned int so that we have 32-bits to work with. The algorithm is going to perform eight shifts to the left so that it can examine each of the eight bits in b. For example, if b is $123 = 01111011_2$ we initialize p with b so that at the start p looks like,

```
                             p
─────────────────────────────────────────────────────
0000 0000 0000 0000 0000 0000 0111 1011
           100s    10s    1s
```

where the bit positions of the hundreds, tens and ones digits for our BCD representation are indicated.

The loop in lines 7 through 24 runs eight times and shifts p one bit position to the left each time. However, before the shift we look at each of the BCD digit locations to see if the value there is five or greater. If it is, we add three before

```
 1| unsigned short bin2bcd8(unsigned char b) {
 2|     unsigned int p=0;
 3|     unsigned char i,t;
 4|
 5|     p = (unsigned int)b;
 6|
 7|     for(i=0; i<8; i++) {
 8|         t = (p & 0xf00) >> 8;
 9|         if (t >= 5) {
10|             t += 3;
11|             p = ((p>>12)<<12)|(t<<8)|(p&0xff);
12|         }
13|         t = (p & 0xf000) >> 12;
14|         if (t >= 5) {
15|             t += 3;
16|             p = ((p>>16)<<16)|(t<<12)|(p&0xfff);
17|         }
18|         t = (p & 0xf0000) >> 16;
19|         if (t >= 5) {
20|             t += 3;
21|             p = ((p>>20)>>20)|(t<<16)|(p^0xffff);
22|         }
23|         p <<= 1;
24|     }
25|
26|     return (unsigned short)(p>>8);
27| }
```

Fig. 2.10 Unsigned 8-bit binary to three digit BCD conversion

shifting to the left. Line 8 pulls out the four bits representing the ones value in the BCD representation. Specifically, it first masks p with 0xf00 which leaves only the four bits in the ones place nonzero. It then shifts to the right eight bits and assigns this value to t. This sets t to the ones value. Line 9 asks if this value is greater than or equal to five. If so, line 10 increments t by three. Line 11 then updates the proper position in p with the new value of t. Let's look at this line more closely. Recall that OR combines bits and here three pieces are combined to replace the proper four bits of p with the value of t. First, ((p>>12)<<12) loses the lower 12 bits of p, which includes the four bits of the BCD ones digit along with the eight bits originally set to b. It then shifts back up so that the net result is a value that has the lowest 12 bits set to zero and the remaining bits set to whatever they were previously. This is OR-ed with t shifted up eight bits so that the lower four bits of t, which are the only ones to have a nonzero value, are in the proper position to be the BCD ones digit. Lastly, p & 0xff keeps the lowest eight bits of p and OR's them in as well. When each of these operations is complete p is the same as it was previously with the exception that the four bits in the BCD ones position have been increased by three. We will see below why three was added and why five was the cutoff value. Lines 13 through 22 perform the same check for the BCD tens and BCD hundreds digits. The only change is the mask and shift values to isolate and work with the correct four bits of p. Lastly, line 23 shifts all of p one position to the left to move the next bit of b (in the lowest eight bits of p) into position. After all bits of b have been examined the final BCD result is in bits nine through twenty so we shift down eight positions to get the final BCD value.

The operation of the entire algorithm is shown below for b = 123. For convenience we ignore the top three nibbles of p, a 32-bit unsigned integer. This leaves us with,

100s	10s	1s	b	comment
0000	0000	0000	0111 1011	initial
0000	0000	0000	1111 0110	shift 1
0000	0000	0001	1110 1100	shift 2
0000	0000	0011	1101 1000	shift 3
0000	0000	0111	1011 0000	shift 4
0000	0000	1010	1011 0000	add 3, ones
0000	0001	0101	0110 0000	shift 5
0000	0001	1000	0110 0000	add 3, ones
0000	0011	0000	1100 0000	shift 6
0000	0110	0001	1000 0000	shift 7
0000	1001	0001	1000 0000	add 3, tens
0001	0010	0011	0000 0000	shift 8

which after the last left shift has 123 as a BCD number in the indicated columns. The return statement shifts this value to the right eight bits to return the final three digit BCD number.

This shows us that the algorithm works but not why it works. The key to understanding the algorithm is to recall that to convert a BCD digit that is not a proper digit, say 1011 which is 11, to a proper digit is to add six, 0110. This will produce a carry bit to the next BCD digit position while preserving the proper value in the existing digit. So, one way to think of how to convert binary to BCD is to look at each digit position and if it is greater than ten subtract ten before doubling. This is to say, calculate,

$$2x + 6$$

for cases where the digit, x, is ten or greater before looking at the next bit of our binary number. This is equivalent to considering if the value is greater than five and if so adding three and then doubling because,

$$2x + 6 = 2(x + 3)$$

which has the added advantage of not requiring an extra bit since $2x + 6$ may be greater than 15 while $x + 3$ never will be for x a valid BCD digit. This is the approach of the double-dabble algorithm shown in Fig. 2.10.

Conversion from a three digit BCD number to binary is especially straightforward. We simply examine each nibble, multiply it by the proper power of 10, and accumulate the result,

```
unsigned char bcd2bin8(unsigned short b) {
    unsigned char n;

    n = (b & 0x0f);
    b >>= 4;
    n += 10*(b & 0x0f);
    b >>= 4;
    n += 100*(b & 0x0f);

    return n;
}
```

The phrase b & 0x0f masks off the lowest nibble of b. This is the ones digit, so we simply set the output value in n to it. We then shift b to the right by four bits to make the lowest nibble the tens digit. We add this to the running total after first multiplying it by ten. We then shift one last time to make the hundreds digit the first nibble, multiply it by 100 and add to n to get the final value of the three digit BCD number.

2.7 Chapter Summary

In this lengthy but essential chapter we covered a lot of important topics. We learned key terminology regarding bits, bytes, and words. We explored unsigned integers at length learning how to perform such low-level functions as manipulation of individual bits to the basic logic operations of AND, OR, NOT and XOR. We completed our tour of low-level manipulation by studying shifts and rotates and how they affect the value of an unsigned integer. We then examined how to compare the magnitude of two unsigned integers.

Unsigned integer arithmetic was our next topic and with it we learned how to implement the basic operations of addition, subtraction, multiplication and division with remainder. To round out the unsigned integers we reviewed one approach for calculating the square root of a number.

Signed integers were our next target. We built upon what we learned with unsigned integers and delved into the main ways signed data is stored: sign-magnitude, one's complement and two's complement. We learned how to compare signed integers and how to either extend existing unsigned arithmetic routines to handle signed arguments or how to implement algorithms that operate on two's complement numbers directly.

To round out our examination we took a look at binary-coded decimal numbers, how to add and subtract them, in ten's complement, and how to convert between binary-coded decimal and pure binary as well as vice versa.

It would not be too much to say that integers are the heart of computers. We now know how to work with them at any level. There are more specialized representations of integers, and we will examine them in other chapters, but for the vast majority of computer work involving data that does not need fractions, including characters which are represented as integer codes, fixed-width integers are our main tools and understanding them in some detail is well worth the effort.

Exercises

2.1. Interpret the following bit patterns first as little-endian then as big-endian unsigned short integers. Convert your answer to decimal.

- 1101011100101110
- 1101010101100101
- 0010101011010101
- 1010101011010111

2.2. Given $A = 6B_{16}$ and $B = D6_{16}$, write the result of each expression in binary and hexadecimal.

- (A AND B) OR (NOT A)
- (A XOR B) OR (A AND (NOT B))
- ((NOT A) OR (NOT B)) XOR (A AND B)

2.3. Using C syntax write an expression that achieves each of the following. Assume that the unsigned char A = 0x8A.

- Set bits 3 and 6 of A.
- Keep only the low nibble of A.
- Set unsigned char B to 100 if bit 3 of A is not set.
- Clear bit 1 of A.
- Toggle bits 4 and 5 of A.

2.4. Using C syntax write an expression for each of the following.

- Swap the upper and lower nibbles of unsigned char v = 0xC4.
- Multiply v by 5 using at least one shift operation.

2.5. Express each of the following numbers in one's and two's complement notation. Write your answer in binary and hexadecimal. Assume 8-bit signed integers.

- -14
- -127
- -1

2.6. Write a function to reverse the bits of an unsigned 8-bit integer. *

2.7. Write a function to count the number of 1 bits in an unsigned 32-bit integer. *

2.8. The *Hamming distance* between two integers is the number of places where their corresponding bits differ. For example, the Hamming distance between 1011 and 0010 is 2 because the numbers differ in bits 0 and 3. Write a function to calculate the Hamming distance between two unsigned short integers. *

2.9. A *Gray code* is a sequence of bit patterns in which any two adjacent patterns in the sequence differ by only one bit. Gray codes were developed to aid in error correction of mechanical switches but are used more generally in digital systems. For example, for four bits, the first six values in the Gray code are: 0000, 0001, 0011, 0010, 0110, 0111. The rule to convert a binary number n into the corresponding Gray code g is to move bit by bit, from lowest bit to highest bit, setting the output bit to the original bit or inverted if the next higher bit is 1. So, to change $n = 0011$ to $g = 0010$ we output a 0 for bit 0 of g since bit 1 of n is set and output the existing bit value of n for all other positions since the next higher bit is not set.

Using this rule, write a C function that calculates the Gray code g for any unsigned short n input. *(Hint: there are many ways to write a function that works but in the end it can be accomplished in one line of code. Think about how to check if the next higher bit of the input is set by shifting and how to invert the output bit if it is.)* **

References

1. Booth, A.D., A signed multiplication technique. The Quarterly Journal of Mechanics and Applied Mathematics, Vol IV, Part 2 (1951)
2. Shannon, C., "A mathematical theory of communication." ACM SIGMOBILE Mobile Computing and Communications Review 5.1 (2001): 3–55.
3. As mentioned in *Anecdotes*, Annals of the History of Computing, Vol 6, No 2, April pp 152–156 (1984)
4. Swift, J., Gulliver's Travels (1726). Reprinted Dover Publications; Unabridged edition (1996)
5. The Internet Protocol. RFC 791 (1981)
6. Boole, G., An Investigation of the Laws of Thought (1854). Reprinted with corrections, Dover Publications, New York, NY, 1958. (reissued by Cambridge University Press, 2009)
7. Gao, S., Al-Khalili, D., Chabini, N., "An improved BCD adder using 6-LUT FPGAs", IEEE 10th International New Circuits and Systems Conference (NEWCAS 2012), pp. 13–16 (2012)

Chapter 3
Floating Point

Abstract We live in a world of floating-point numbers and we make frequent use of floating-point numbers when working with computers. In this chapter we dive deeply into how floating-point numbers are represented in a computer. We briefly review the distinction between real numbers and floating-point numbers. Then comes a brief historical look at the development of floating-point numbers. After this we compare two popular floating-point representations and then focus exclusively on the IEEE 754 standard. For IEEE 754 we consider representation, rounding, arithmetic, exception handling, and hardware implementations. We conclude the chapter with some comments about the pitfalls associated with using floating-point numbers.

3.1 Floating-Point Numbers

This chapter is about floating-point numbers. This is a highly complex area with active research even today. As such, we can only scratch the surface of representation and computation with floating-point numbers in a single chapter. Therefore, we will, after some preliminaries, focus on the current Institute of Electrical and Electronics Engineers (IEEE) 754-2008 standard which we will refer to simply as "IEEE 754". For a more in-depth treatment of floating-point numbers and arithmetic please see Muller et al. [1].

It would be good to have an idea of what, exactly, a "floating-point" number actually is and how it differs from a "real" number. Let's start with the latter concept. This one we know from school: a real number is a number that can be found on the real number line. We have an intuitive feel for what a real number is, and we know many of them because we know that all integers ($\mathbb{Z}$) and rationals ($\mathbb{Q}$) are also real numbers. But, of course, there are still many more real numbers besides these. For example, π is transcendental and therefore a real number but it is neither an integer nor a rational (capable of being written as p/q with $p, q \in \mathbb{Z}$).

© Springer International Publishing Switzerland 2015
R.T. Kneusel, *Numbers and Computers*, DOI 10.1007/978-3-319-17260-6_3

How, then, to define the real numbers, $\mathbb{R}$? Formally, there are several ways to define $\mathbb{R}$. The one we will consider is the synthetic approach with specifies $\mathbb{R}$ by defining several axioms which hold only for $\mathbb{R}$. The set of numbers which satisfy these axioms is, by definition, $\mathbb{R}$. Specifically, $\mathbb{R}$ may be defined by considering a set ($\mathbb{R}$), two special elements of $\mathbb{R}$, 0 and 1, two operations called addition ($+$) and multiplication ($\times$), and finally a relation on $\mathbb{R}$ called $\leq$ which orders the elements of $\mathbb{R}$. With this and the following four axioms, we have arrived at a specification of the real numbers. The axioms are,

1. $(\mathbb{R}, +, \times)$ forms a field.
2. $(\mathbb{R}, \leq)$ forms a totally ordered set.
3. $+$ and $\times$ maintain $\leq$.
4. The order $\leq$ is complete in a formal sense.

where a full description of their meaning and implications are beyond the scope of this text. The interested reader is directed to [2] for a presentations of fields and rings. For the present, in this sense, $+$ is simply normal addition and $\times$ is normal multiplication. The order $\leq$ also acts as expected. A *floating-point number* is an approximation of a real number that is representable on a computer. It is "close enough" (we'll see below what that means) to the actual real number we wish to represent to be useful and lead to meaningful results.

To be specific, in this chapter a floating-point number is a number which is represented in exponential form as,

$$\pm d_0.d_1 d_2 d_3 \ldots d_{p-1} \times \beta^e, \ 0 \leq d_i < \beta$$

where $d_0.d_1 d_2 d_3 \ldots d_{p-1}$ is the p digit *significand* (or *mantissa*), β is the base and e is the integer exponent. For the above we follow the notation given in [3]. While it is possible for the base β to be any integer for this chapter we restrict ourselves to $\beta = 2$ so that our numbers are binary and the only allowed digits are 0 and 1.

If there are p digits in the significand then this part of the floating-point number can have β^p unique values. This defines the precision with which values can be represented for a particular exponent. To complete the format we need to also specify the range for the exponent from e_{min} to e_{max}. Before moving on to a brief history of floating-point numbers in computers let's take a quick look at working with floating-point numbers in base 2.

The significand, $d_0.d_1 d_2 d_3 \ldots d_{p-1}$ is a number with the value of

$$d_0 + d_1 \beta^{-1} + d_2 \beta^{-2} + d_3 \beta^{-3} + \cdots + d_{p-1} \beta^{-(p-1)}$$

which is then scaled by the exponent β^e, so that the final value of the number is,

$$(d_0 + d_1 \beta^{-1} + d_2 \beta^{-2} + d_3 \beta^{-3} + \cdots + d_{p-1} \beta^{-(p-1)}) \beta^e$$

where we have ignored a possible minus sign to make the entire number negative.

The above makes it clear that it is the number of digits p in the significand that controls the precision of the floating-point number for a given exponent e. It also means that the distribution of floating-point numbers on the real number line is not uniform as shown in Fig. 3.1.

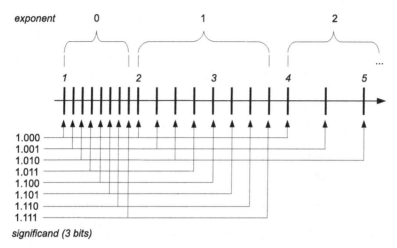

Fig. 3.1 The distribution of floating-point numbers. For this figure the number of significand bits is three with an implied leading bit of one: $1.m_0m_1m_2 \times 2^e$. The location of representable numbers for exponents of $e = 0, 1, 2$ are shown. For each increase in the exponent the number of representable numbers is halved. So, between $[1, 2)$ there are eight values while between $[2, 3)$ there are only four and between $[4, 5)$ there are only two

This is the case because as we saw above the digits of the significand become multipliers on the exponent value β^e. So, for a fixed exponent, the smallest interval between two floating-point numbers is $\beta^{-(p-1)}\beta^e$ which is of fixed size as long as the exponent does not change. Once the exponent changes from e to $e + 1$ the minimum interval becomes,

$$\beta^{-(p-1)}\beta^e \rightarrow \beta^{-(p-1)}\beta^{e+1} = \left(\beta^{-(p-1)}\beta^e\right)\beta$$

which is β times larger than the previous smallest interval.

Above we said that there are β^p possible significands. However, not all of them are useful if we want to make floating-point numbers unique. For example, in decimal 1.0×10^2 is the same as 0.1×10^3 and the same holds true for any base β. Therefore, to make floating-point numbers uniquely representable we add a condition which states that the first digit of the significand must be nonzero. For binary floating-point this means that the first digit of the significand must be a 1. Floating-point numbers with this convention are said to be *normalized*. This requirement also buys us one extra digit of precision when the base is 2 because the only nonzero digit is 1 which means all normal binary floating-point numbers have a known leading digit for the significand. This known digit need not be stored in memory leaving room for a $p + 1$-th digit in the significand.

To make all of this concrete, let's look at expressing some binary floating-point numbers ($\beta = 2$). We will calculate the actual decimal values for the floating-point numbers shown in Fig. 3.1. They are,

significand	exponent	expanded	decimal value
1.000	0	$(1) \times 1$	1.0000
1.001	0	$(1 + \frac{1}{8}) \times 1$	1.1250
1.010	0	$(1 + \frac{1}{4}) \times 1$	1.2500
1.011	0	$(1 + \frac{1}{4} + \frac{1}{8}) \times 1$	1.3750
1.100	0	$(1 + \frac{1}{2}) \times 1$	1.5000
1.101	0	$(1 + \frac{1}{2} + \frac{1}{8}) \times 1$	1.6250
1.110	0	$(1 + \frac{1}{2} + \frac{1}{4}) \times 1$	1.7500
1.111	0	$(1 + \frac{1}{2} + \frac{1}{4} + \frac{1}{8}) \times 1$	1.8750
1.000	1	$(1) \times 2$	2.0000
1.001	1	$(1 + \frac{1}{8}) \times 2$	2.2500
1.010	1	$(1 + \frac{1}{4}) \times 2$	2.5000
1.011	1	$(1 + \frac{1}{4} + \frac{1}{8}) \times 2$	2.7500
1.100	1	$(1 + \frac{1}{2}) \times 2$	3.0000
1.101	1	$(1 + \frac{1}{2} + \frac{1}{8}) \times 2$	3.2500
1.110	1	$(1 + \frac{1}{2} + \frac{1}{4}) \times 2$	3.5000
1.111	1	$(1 + \frac{1}{2} + \frac{1}{4} + \frac{1}{8}) \times 2$	3.7500
1.000	2	$(1) \times 4$	4.0000
1.001	2	$(1 + \frac{1}{8}) \times 4$	4.5000
1.010	2	$(1 + \frac{1}{4}) \times 4$	5.0000

where we see that the three bits of the significand has indeed set the precision of each number for a fixed exponent but that the larger the exponent the greater the delta between successive floating-point numbers.

3.2 An Exceedingly Brief History of Floating-Point Numbers

Perhaps the first description of floating-point numbers, meaning a method for storing and manipulating real numbers with a machine, is that of Leonardo Torres y Quevedo in his paper "Essays on Automatics" (1914). This paper is included in Randell's book "The Origins of Digital Computers" [4]. Torres y Quevedo implemented an electromechanical calculating machine, the "electromechanical arithmometer", inspired by the work of Charles Babbage, which was capable of accepting typed statements such as *365 x 256* via a typewriter and would respond, by typing on the same line, with = *93440* and then advance to the next line. This machine was demonstrated at a conference in Paris in 1920.

As for floating-point, Torres describes a way to store floating-point numbers in a consistent manner in his "Essays" paper. First he states that numbers will be stored in exponential format as $n \times 10^m$. He then offers three rules by which consistent manipulation of floating-point numbers by a machine could be implemented. They are, in his own words (as translated into English),

1. *n will always be the same number of digits (six for example).*
2. *The first digit of n will be of order of tenths, the second of hundredths, etc.*
3. *One will write each quantity in the form: n;m.*

On the whole this is quite remarkable for 1914. The format he proposed shows that he understood the need for a fixed-sized significand as is presently used for floating-point data. He also fixed the location of the decimal point in the significand so that each representation was unique. Finally, he even showed concern for how to format such numbers by specifying a syntax to be used that could be entered through a typewriter, as was the case for his "electromechanical arithmometer" some years later. We use essentially the same syntax for most programming languages today; simply replace the ";" with an "E" or "e" so that where he would have written *0.314159;1* we now write *0.314159E1*. It is unfortunate that Torres y Quevedo's work is not more widely known outside his native Spain.

Konrad Zuse is typically credited with being first to implement practical floating-point numbers for his *Z1* and *Z3* computers [5]. The *Z1* was developed in 1938 with the *Z3* completed in 1941. These computers used binary floating-point numbers with one bit for the sign, 14 bits for the significand, and a signed exponent of 7 bits. Zuse also recognized the value of an implicit "1" bit in the significand and stored the numbers in normalized format. This format is in many ways identical to the modern floating-point formats we will be investigating in this chapter and is a testament to Zuse's often unrecognized brilliance.

According to [5] Zuse called any number with an exponent of -64 "zero" and any number with an exponent of 63 "infinity". This means that the smallest number possible on the *Z3* was $2^{-63} = 1.08 \times 10^{-19}$ while the largest was $1.999 \times 2^{62} = 9.2 \times 10^{18}$. Numbers were entered in decimal and converted to binary format. Similarly, output was converted from binary to decimal. However, this conversion did not cover the entire range of the machine and the user was limited to a much smaller floating-point range for input and output.

Computers in the 1950s through the 1970s used a wide variety of floating-point formats. For example, the Burroughs B6700 used a base 8 ($\beta = 8$) format while the IBM System/360 used a base 16 ($\beta = 16$) format. Many other computers copied the IBM format including the Data General Nova and Eclipse computers.

In 1985 IEEE released a standard for floating-point arithmetic. This standard has since dominated the industry and has been implemented in hardware by the Intel x86 series of microprocessors. This standard is the main focus of this chapter and we begin by comparing its floating-point representation to the older IBM format.

3.3 Comparing Floating-Point Representations

In this section we examine the 32-bit IBM S/360 format for floating-point numbers and compare it to the 32-bit IEEE 754 format. We then focus exclusively on the IEEE format for the remainder of the section and chapter. By this comparison we will see that the choice of $\beta = 16$ was not a particularly good one and why the $\beta = 2$ of IEEE is a better one.

A single-precision floating-point number in the IBM S/360 format was stored in memory as,

with 1 bit for the sign (S), 7 bits for the exponent (E), and 24 bits for the significand (M). The exponent was stored in excess-64 format meaning an exponent of 0 represented -64 and an exponent of 127 represented 63. The significand was in binary but manipulated in groups of four bits since the base was 16 and $2^4 = 16$. Because of this, one can think of the significand as representing six hexadecimal digits for the first six negative powers of 16.

In order to normalize a floating-point number at least one of the first four high bits of the significand had to be nonzero. Because of this the smallest positive floating-point number that could be represented was,

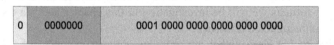

which is, in base 16, $0.1_{16} \times 16^{0-64} = 1 \times 16^{-1} \times 16^{-64} = 5.397605 \times 10^{-79}$.

Similarly, the largest positive number was,

equal to $0.\text{FFFFFF}_{16} \times 16^{127-64} = 15 \times 16^{-1} + 15 \times 16^{-2} + 15 \times 16^{-3} + 15 \times 16^{-4} + 15 \times 16^{-5} + 15 \times 16^{-6} \times 16^{63} = (1 - 16^{-6}) \times 16^{63} = 7.237005 \times 10^{75}$ since the significand of $0.\text{FFFFFF}_{16}$ is only one 16^{-6} away from 1, hence $0.\text{FFFFFF}_{16} = 1 - 16^{-6}$. The number zero was represented by all bits set to zero.

The IEEE 754 standard defines a number of different floating-point sizes. For the moment we consider only the *binary32* single-precision format which fits in a 32-bit number. On modern x86-based hardware this is the format used when one declares a variable to be of type float in C, for example. This format is stored in memory as, with 1 bit for the sign (S), 8 bits for the exponent (E), and 23 bits for the significand (M). The exponent is stored in excess-127 format with two special exponent values of 0 and FF_{16} reserved. The base is 2 ($\beta = 2$) and the significand has an implied leading 1 bit always for a normalized number. So, in this format, the number stored is,

$$\pm 1.b_{22}b_{21}b_{20} \ldots b_0 \times 2^{E-127}$$

where b_n is the value of that bit of the significand. Since the base is 2 the bits are interpreted as individual bits as opposed to groups of four bits for the base 16 IBM format.

Because of the implied 1 bit, the smallest allowed *binary32* number is,

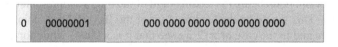

which is $1 \times 2^{1-127} = 1.1754944 \times 10^{-38}$. Similarly, the largest positive number is,

which, using the reasoning above for the significand of the largest IBM S/360 format number, is $(2 - 2^{-23}) \times 2^{254-127} = 3.4028235 \times 10^{38}$. The mantissa is $(2 - 2^{-23})$ instead of $(1 - 2^{-23})$ because of the implied leading 1 bit of the *binary32* significand.

Immediately we see that the IBM format has a much larger range than the IEEE format, approximately $[10^{-79}, 10^{75}]$ versus $[10^{-38}, 10^{38}]$. This is encouraging in terms of expressing large numbers using fewer bits of memory but there is a problem hidden in the significand. Namely, for the IBM format, a normalized number is one in which the first hexadecimal digit of the six hexadecimal digit significand is non-zero. This means that any of the possible bit patterns for 1 through 15 are allowed. If the bit pattern is 0001_2 then there are three fewer significant bits in the significand than if the leading digit is 1000_2 up to 1111_2. This represents a potential lost of precision of up to one decimal digit and is due entirely to the way in which the number is represented in base 16. This loss of precision as a function of the representation of the number in the selected base is known as *wobbling precision* and is more severe when the base is larger, as it is for the IBM format. This wobbling precision can lead to very inaccurate results as the following example illustrates (borrowed from Cody and Waite [6]).

The approximate value of $\pi/2$ in hexadecimal is $1.922_{16} \times 16^0$. We get this from Python using the `float.hex()` class method:

$$\text{float.hex(pi/2.0)} = \text{0x1.921fb54442d18p+0}$$
$$= 1.921\text{fb}54442\text{d}18_{16} \times 2^0$$

where we must be careful because Python expresses floats in hexadecimal with a base of 16, not 2. The conversion to binary is straightforward, however, simply replace each hexadecimal digit with four binary digits equal to the hexadecimal digit. In this case, the exponent is zero so the actual value is just the mantissa. Similarly, the approximate value of $2/\pi$ in hexadecimal is $\text{a.2f9}_{16} \times 16^{-1}$ since,

$$\text{float.hex(2.0/pi)} = \text{0x1.45f306dc9c883p-1}$$
$$= 1.45\text{f}306\text{dc}9\text{c}883_{16} \times 2^{-1}$$
$$= 1.45\text{f}306\text{dc}9\text{c}883_{16}/2$$
$$\approx \text{a.2f9}_{16} \times 16^{-1}$$

with the important point being the leading digit of the base 16 representation of these two numbers. This would be the leading digit of the IBM floating-point format for these numbers and in the case of $\pi/2$ we see that it is a 0001_2 while for $2/\pi$ we get 1010_2. The former has three leading zeros meaning only 21 significant bits in the IBM representation while the latter has no leading zeros meaning all 24 significand bits are important. This difference of three bits is a loss of about one decimal digit of accuracy.

What is the case for the IEEE base 2 format? In this case there is no loss of precision at all which we can see immediately from the Python output. For $\pi/2$ we get $1.921\text{fb}54442\text{d}18_{16} \times 2^0$ while for $2/\pi$ we have $1.45\text{f}306\text{dc}9\text{c}883_{16} \times 2^{-1}$ both of which have a leading 1 digit as the IEEE format specifies. This means that all the bits of the significand are meaningful and there is no wobbling precision effect when $\beta = 2$. This is the reason why the IEEE 754 format is to be preferred to the older IBM S/360 floating-point format.

As an aside, before we leave this section, we describe a small C function for converting 32-bit IBM S/360 floating point numbers to IEEE double precision numbers. Double precision is necessary as we have seen that the single precision IEEE format does not have the numeric range necessary to hold all 32-bit IBM floats. The IBM format is assumed to be stored in an unsigned 32-bit integer which allows all bit patterns. The function is,

```
 1  double ibm_to_ieee_64(uint32_t n) {
 2      int s, e, m;
 3      double ans;
 4
 5      s = (n & (1<<31)) != 0;
 6      e = (n >> 24) & 0x7F;
 7      m = n & 0xFFFFFF;
 8
 9      ans = m * pow(16,-6) * pow(16,e-64);
10      return (s == 1) ? -ans : ans;
11  }
```

This function may be useful if one encounters an old data file in IBM format (this is also the format used by many other computer systems in the 1970s and 1980s). The sign of the IBM float is extracted in line 5, the exponent in line 6, and the significand in line 7 using simple masking and shifting operations. The resulting double precision floating point number is calculated in line 9. The significand of the IBM format is of the form $0.h_0h_1h_2h_3h_4h_5$ so the value in m must be multiplied by 16^{-6} before the final multiplication by 16^{e-64} where we subtract 64 because the exponent is in excess-64 format. Line 10 simply sets the sign of the output and returns it. This function assumes that it is running on a system which uses IEEE floating-point by default.

3.4 IEEE 754 Floating-Point Representations

The IEEE 754 standard defines multiple floating-point formats for binary ($\beta = 2$) and decimal ($\beta = 10$) numbers. We are only concerned with the binary formats here. These are, with their ranges,

Name	Minimum	Maximum
binary16	$2^{-14} \approx 6.104 \times 10^{-5}$	$(2-2^{-10}) \times 2^{15} = 65504.0$
binary32	$2^{-126} \approx 1.175 \times 10^{-38}$	$(2-2^{-23}) \times 2^{127} \approx 3.403 \times 10^{38}$
binary64	$2^{-1022} \approx 2.225 \times 10^{-308}$	$(2-2^{-52}) \times 2^{1023} \approx 1.798 \times 10^{308}$

where *binary16* is half precision, *binary32* is single precision (C `float`), *binary64* is double precision (C `double`), and *binary128* (not shown) is quadruple precision, sometimes called "quad precision". The *binary128* format uses 128-bits and some C compilers will implement it in this format using software. More often, however, compilers for x86 architectures will use the 80-bit extended precision hardware type instead. Because of this we will largely ignore *binary128* in this chapter. If you are familiar with Fortran, the *binary32* format is roughly equivalent to a `REAL`, while the *binary64* format is similar to `REAL*8` and the *binary128* format is like `REAL*16`.

The *binary32* format was illustrated above in Sect. 3.3. For completeness we illustrate *binary16*,

and *binary64*,

which follow the same format as *binary32* but differ in the exponent and significand bits. For *binary16* the exponent is 5 bits and is stored in excess-15 format. The exponent for *binary32* is 8 bits and uses excess-127 format. Lastly, *binary64* uses an exponent of 11 bits stored in excess-1023 format.

The observant reader will see that the exponent ranges in the table above do not cover the full range of bit values possible. For example, the *binary32* format uses 8 bits for the exponent which means any exponent from $0 - 127 = -127$ to $255 - 127 = 128$ should be possible but the actual exponent range is from -126 to 127 instead. This is because IEEE 754 uses the smallest and largest exponent values for specific purposes. The smallest exponent value, the bit pattern of all zeros, designates either zero itself, or a *subnormal* number. We will deal with subnormal numbers shortly. The largest exponent bit pattern, either 11111_2, 11111111_2, or 11111111111_2 for *binary16*, *binary32*, or *binary64*, respectively, is reserved for infinity or NaN (not-a-number). The selection between zero and a subnormal number or infinity or NaN depends upon the value of the significand (M). Specifically,

Exponent	$M = 0$	$M \neq 0$
largest	$\pm\infty$	NaN
smallest (=0)	± 0	*subnormal*

where the sign bit sets positive or negative. Note carefully that this convention implies that zero itself is signed but the standard requires $+0 = -0$, i.e., zero is interpreted as zero, regardless of the sign bit.

Most readers are likely familiar with the single and double precision floating point numbers, *binary32* and *binary64*, respectively. These generally map to the well used C data types of `float` and `double`. Note, however, that the Python data type `float` is really a *binary64* number. The addition of *binary16* to the IEEE standard seems a bit odd as it does not map to any primitive data type typically used in modern programming languages.

Half precision floats, another common name for *binary16* numbers, are supported in recent versions of `gcc` for ARM targets. Newer Intel microprocessors also have instructions for working with *binary16* numbers. In both cases the storage in memory uses half precision but once read from memory the numbers are converted to 32-bit floats. The main use for these numbers is to increase the precision of values that are stored on graphics processors. Indeed, *OpenGL* supports *binary16* numbers. Naturally, however, this comes with a price, which is a rapidly increasing loss of precision as the value stored gets further and further from zero. As illustrated in Fig. 3.1 the difference between floating-point values increases as one moves further away from zero. On the plus side, this means that if one needs to work with numbers in the range $[0, 1]$ then unless very high precision is required *binary16* is a reasonable way to keep relatively high precision (about 6 or 7 decimals) while using half the memory of a 32-bit float and one quarter the memory of a 64-bit float. This might be significant if there are large arrays of these values to work with as the bandwidth to move that memory is much less than it might otherwise be. It also may help by allowing more data to fit in a processor cache thereby reducing the number of expensive and slow accesses from main memory. Recall that images are typically stored as large arrays so it makes sense to want to be as efficient as possible with memory while not sacrificing precision or dynamic range.

Infinity For any of the IEEE formats, if the exponent is maximum, i.e., all bits set to 1, and the significand is zero, then the number is infinite (∞) and positive or negative depending upon the sign bit. Infinity is produced when the result of an operation would exceed the range of numbers that can be expressed in the given floating-point format. For example,

$$\begin{aligned} \text{exp}(1000000.0) &= \text{Inf} = +\infty \\ -\text{exp}(1000000.0) &= -\text{Inf} = -\infty \end{aligned}$$

Infinity is also returned for certain other operations such as,

$$\begin{aligned} 1.0/0.0 &= \text{Inf} = +\infty \\ \log(0.0) &= -\text{Inf} = -\infty \end{aligned}$$

which follow the usual rules of arithmetic. The $-\infty$ result for $\log(0.0)$ makes sense as a limit as $x \to 0$ from the right.

As mentioned in the IEEE standard, the following operations are valid and work as expected when using ∞ as an argument and x finite (and not zero),

$$\begin{aligned} \infty + x &= \infty \,, & x + \infty &= \infty \\ \infty - x &= \infty \,, & x - \infty &= -\infty \\ \infty \times x &= \infty \,, & x \times \infty &= \infty \\ \infty \div x &= \infty \,, & x \div \infty &= 0 \end{aligned}$$

Additionally, the square root of infinity is also defined to be infinity.

Not-A-Number While infinity is straightforward, Not-a-Number or NaN, gets a little messy. A NaN is stored as a floating-point number with the largest possible exponent (all bits 1) and a significand that is *not* zero. There are two kinds of NaNs, quiet and signaling. We will discuss this distinction more fully below when we talk about floating-point exceptions. For now we are concerned with how these two kinds of NaNs are stored so that we can distinguish between them.

According to the standard, the first bit of the significand, which is d_1 in the representation $\pm d_0.d_1 d_2 d_3 \ldots d_{p-1} \times \beta^e$ since d_0 is always implied and 1, determines the NaN type. If this bit is zero the NaN is a signaling NaN, otherwise it is a quiet NaN. Note that this implies that a signaling NaN must have at least one other bit of the significand be nonzero otherwise the number is not a NaN but infinity. The remaining $p - 2$ digits of the significand are available for a "payload", to use the standard's term, which may contain diagnostic information about what caused the NaN in the first place.

  The standard is intentionally fuzzy about what, if anything, should constitute the payload of a NaN. However, the standard is explicit that NaNs, when used as an argument in an operation, should preserve the payload of at least one of the NaNs, or the NaN, if it is the only NaN in the operation. Operations on NaNs produce NaNs as output. The vague nature of the NaN payload invites creative use. Indeed, some groups and companies have used NaNs for their own purposes. Since it is not specified, beyond indicating a bit for quiet or signaling, users are free to use NaNs as they see fit. One possibility might be to use NaNs as symbols in a symbolic expression parser. Another would be to use NaNs as missing data values and the payload to indicate a source for the missing data or its class.

  The following set of C functions can be used to check for NaN and to set and get a payload from a NaN of up to 22 bits,

```
 1 typedef union {
 2     float f;
 3     unsigned int d;
 4 } fp_t;
 5
 6 char nan_isnan(float nan) {
 7     fp_t x;
 8     x.f = nan;
 9     return (((x.d >> 23) & 0xFF) == 0xFF) && ((x.d & 0x7FFFFF) != 0);
10 }
11
12 float nan_set_payload(float nan, unsigned int payload) {
13     fp_t x;
14     x.f = nan;
15     x.d |= (payload & 0x3FFFFF);
16     return x.f;
17 }
18
19 unsigned int nan_get_payload(float nan) {
20     fp_t x;
21     x.f = nan;
22     return (x.d & 0x3FFFFF);
23 }
```

where nan_isnan checks to see if a float is actually a NaN, nan_set_payload sets the free 22-bits of a NaN to a user-specified value and nan_get_payload returns this value from a NaN. Most systems already have a isnan function, we include it here for completeness. The functions make use of a trick to work with the same piece of memory as both a float and an unsigned integer, both of which occupy 32-bits. This is the fp_t data type defined in lines 1 through 4. The function nan_isnan accepts a *binary32* number and puts it into x (line 8). We then interpret the memory of the float as an unsigned integer by accessing the d field. Line 9 is the magic in two parts both of which come directly from the definition of a NaN. The first part, (((x.d >> 23) & 0xFF) == 0xFF), examines the exponent field of the number. Recall that a *binary32* number has 23 bits in the significand. We shift 23 bits down to move the exponent field into bit position 0. We then mask this to keep the first eight bits (the AND with 0xFF). This removes any sign bit which the standard ignores for NaNs. If this is the exponent of an actual NaN then, by definition, the exponent is all 1 bits, which is 0xFF so we check if the value is equal to 0xFF. An exponent of all 1 bits is not enough to define a NaN, however, as the number could still be infinity. We must consider the significand itself. This is the second part of line 9, ((x.d & 0x7FFFFF) != 0), which masks off the 23-bits of the significand by the AND with 0x7FFFFF and then checks to see if this value is zero. If the number is a NaN, this value will not be zero. If it is a quiet NaN the highest bit will be set, at a minimum, and if it is a signaling NaN, at least one of the other 22-bits will need to be set. If both of these parts are true, the number is a NaN and the result of the final logical AND is returned.

The nan_set_payload function accepts a NaN and a payload which must fit in 22-bits. A fp_t is used to hold the NaN. The payload is masked with 0x3FFFFF to keep it in range and then OR'ed into the bits of the NaN in line 15. The updated NaN is returned in line 16. To access the NaN value use nan_get_payload which accepts a NaN and after setting x to it, returns the lowest 22-bits by again masking with 0x3FFFFF.

To illustrate the idea of NaN propagation consider the following code snippet, compiled on a 32-bit Linux system using gcc,

```
 1 const char *pp(unsigned int x) {
 2     static char b[33];
 3     unsigned int z;
 4     b[0] = '\0';
 5     for (z = (1<<31); z > 0; z >>= 1)
 6         strcat(b, ((x & z) == z) ? "1" : "0");
 7     return b;
 8 }
 9
10 int main() {
11     fp_t x;
12     float n = log(-1);
13     unsigned int d;
14
15     x.f = n;
16     printf("%d  original    = %s\n", nan_isnan(n),pp(x.d));
17     n = nan_set_payload(n, 12345);
18
19     x.f = n;
20     printf("%d  set payload = %s\n", nan_isnan(n),pp(x.d));
21     d = nan_get_payload(n);
22     printf("payload is %d (%s)\n", d, pp(d));
23
24     n = 123 * n;
25
26     d = nan_get_payload(n);
27     printf("payload is still %d (%s)\n", d, pp(d));
28 }
```

where the function pp simply prints the given 32-bit unsigned integer in binary as a string. In line 12 we define n and set it so $\log(-1)$ which is a NaN. In line 15 we put this NaN into x and print the result of asking whether n is a NaN (it is) and showing the bits of n,

```
1  original    = 01111111110000000000000000000000
```

Line 17 sets the payload for the NaN to the unimaginative air-shield combination number of 12345. Lines 19 through 22 show that the NaN is still a NaN but now has a payload,

```
1  set payload = 01111111110000000011000000111001
   payload is 12345 (00000000000000000011000000111001)
```

where $11000000111001_2 = 12345$. So far, this isn't anything particularly impressive. Line 24, however, multiplies n by 123 and assigns back to n. This is to illustrate NaN propagation, which the IEEE standard strongly encourages. The output of line 27 is,

```
payload is still 12345 (00000000000000000011000000111001)
```

Which clearly demonstrates that the payload was propagated.

Subnormal Numbers As we previously stated, a normal number is one in which the leading bit of the significand is an implied 1 and the exponent, e, is $0 < e < m$, for m the maximum exponent value which depends upon the precision of the number. For *binary32*, $m = 255$ since eight bits are used to store the exponent. Therefore, the smallest normal *binary32* number is,

$$0\ 00000001\ 00000000000000000000000_2 = 2^{-126} = 1.1754944 \times 10^{-38}$$

where we have separated the binary representation into the sign, exponent and significand. If we insist on using an implied leading 1 bit for the significand the next smallest number we can represent is immediately zero. However, if we are willing to suffer a loss of precision in the significant we can drop the exponent to zero and change the implied leading digit of the significand from 1 to 0 to represent numbers as,

$$\pm 0.d_1 d_2 \ldots d_{23} \times 2^{-126}$$

where numbers of this form are called *subnormal* or *denormal*. Subnormal numbers allow us to work with very small numbers around zero but at the small cost of a loss of precision in the significand. However, there is a very large cost in terms of performance as modern floating-point hardware does not work with subnormal numbers implying calculations must be done in software.

The program in Fig. 3.2 illustrates the transition between normal and subnormal numbers for 32-bit floats. It initializes a float to 1.0 and then repeatedly divides it by 2 until the result is zero. The function ppx prints the bits of the floating point number with a space between the sign and exponent and the exponent and the significand. If we run this program the tail end of the output is,

```
x = (0 00000010 00000000000000000000000) 2.3509887e-38
x = (0 00000001 00000000000000000000000) 1.1754944e-38
x = (0 00000000 10000000000000000000000) 5.8774718e-39
x = (0 00000000 01000000000000000000000) 2.9387359e-39
x = (0 00000000 00100000000000000000000) 1.4693679e-39
x = (0 00000000 00010000000000000000000) 7.3468397e-40
x = (0 00000000 00001000000000000000000) 3.6734198e-40
x = (0 00000000 00000100000000000000000) 1.8367099e-40
x = (0 00000000 00000010000000000000000) 9.1835496e-41
x = (0 00000000 00000001000000000000000) 4.5917748e-41
x = (0 00000000 00000000100000000000000) 2.2958874e-41
x = (0 00000000 00000000010000000000000) 1.1479437e-41
x = (0 00000000 00000000001000000000000) 5.7397185e-42
x = (0 00000000 00000000000100000000000) 2.8698593e-42
x = (0 00000000 00000000000010000000000) 1.4349296e-42
x = (0 00000000 00000000000001000000000) 7.1746481e-43
x = (0 00000000 00000000000000100000000) 3.5873241e-43
x = (0 00000000 00000000000000010000000) 1.793662e-43
x = (0 00000000 00000000000000001000000) 8.9683102e-44
x = (0 00000000 00000000000000000100000) 4.4841551e-44
x = (0 00000000 00000000000000000010000) 2.2420775e-44
x = (0 00000000 00000000000000000001000) 1.1210388e-44
x = (0 00000000 00000000000000000000100) 5.6051939e-45
x = (0 00000000 00000000000000000000010) 2.8025969e-45
x = (0 00000000 00000000000000000000001) 1.4012985e-45
x = (0 00000000 00000000000000000000000) 0
```

where the first two numbers are normalized, the exponent is nonzero. After these start the subnormalized numbers the first of which is,

$$0\ 00000000\ 10000000000000000000000_2 = 2^{-1} \times 2^{-126} = 5.8774718 \times 10^{-39}$$

and the smallest positive subnormal number is therefore $2^{-23} \times 2^{-126} = 1.4012985 \times 10^{-45}$.

The penalty in performance for using subnormal numbers can be found by timing how long it takes to run 10 million iterations of a simple multiplication of x by 1 where x is first the smallest normal number and then the largest subnormal number. If we do this, being careful to make sure the loop is actually performed and not optimized away by the compiler, we see that subnormal numbers are about 23 times slower than normal numbers on the machine used. This reflects the difference between hardware and software support.

3.5 Rounding Floating-Point Numbers (IEEE 754)

Often when a real number is expressed in floating-point format it is necessary to round it so that it can be stored as a valid floating-point number. The IEEE standard defines four rounding modes, or rules, for binary floating-point. These are available to programmers in C on Linux if using the gcc compiler (among others). The rounding modes are listed in Fig. 3.3.

```
 1 |#include <stdio.h>
 2 |#include <string.h>
 3 |
 4 |const char *ppx(unsigned int x) {
 5 |    static char b[36];
 6 |    unsigned int z, i=0;
 7 |    b[0] = '\0';
 8 |    for (z = (1<<31); z > 0; z >>= 1) {
 9 |        strcat(b, ((x & z) == z) ? "1" : "0");
10 |        if ((i == 0) || (i == 8))
11 |            strcat(b, " ");
12 |        i++;
13 |    }
14 |    return b;
15 |}
16 |
17 |typedef union {
18 |    float f;
19 |    unsigned int d;
20 |} fp_t;
21 |
22 |int main() {
23 |    fp_t x;
24 |    int i;
25 |    x.f = 1.0;
26 |    for(i=0; i < 151; i++) {
27 |        printf("x = (%s) %0.8g\n", ppx(x.d), x.f);
28 |        x.f /= 2.0;
29 |    }
30 |}
```

Fig. 3.2 A program to display the transition from normal to subnormal floating-point numbers

Round to Nearest	Round to the nearest floating-point number. Ties round to even.
Round Towards Zero	Round down, if positive, or up, if negative, always towards zero.
Round Towards $+\infty$	Round up, always towards positive infinity.
Round Towards $-\infty$	Round down, always towards negative infinity.

Fig. 3.3 The four IEEE 754 binary floating-point rounding modes

The default mode for gcc, and the one recommended by the IEEE standard, is to round towards the nearest value with ties rounding towards the even value. The even value is the one with a zero in the lowest order bit of the significand. For example, using decimal and rounding to an integer value "round to nearest" would act in this way,

$$+123.7 \rightarrow +124$$
$$+123.4 \rightarrow +123$$
$$+123.5 \rightarrow +124$$
$$-123.7 \rightarrow -124$$
$$-123.4 \rightarrow -123$$
$$-123.5 \rightarrow -124$$

which is more or less the way we would expect rounding to work. The stipulation of rounding ties to even helps prevent the accumulation of rounding errors as in binary this will happen about 50 % of the time.

The other, directed, rounding modes work in this way,

Value	Round towards Zero	Round towards $+\infty$	Round towards $-\infty$
+123.7	+123	+124	+123
+123.4	+123	+124	+123
+123.5	+123	+124	+123
−123.7	−123	−123	−124
−123.4	−123	−123	−124
−123.5	−123	−123	−124

The directed rounding modes always shift results, consistently up or down for rounding towards $\pm\infty$, regardless of sign, or up or down for negative and positive numbers in the case of "round towards zero". The default mode is used so often that many software libraries do not function properly if one of the directed modes is used [7].

It is possible to alter the rounding mode programmatically. Figure 3.4 is a Linux program that changes the rounding mode while displaying the floating-point representation of exp(1.1) and -exp(1.1).

This program produces the following output,

```
FE_TONEAREST:
   exp(1.1)  = (0 10000000 10000000100010001000010) 3.00416613
  -exp(1.1)  = (1 10000000 10000000100010001000010) -3.00416613

FE_UPWARD:
   exp(1.1)  = (0 10000000 10000000100010001000010) 3.00416613
  -exp(1.1)  = (1 10000000 10000000100010001000001) -3.00416589

FE_DOWNWARD:
   exp(1.1)  = (0 10000000 10000000100010001000001) 3.00416589
  -exp(1.1)  = (1 10000000 10000000100010001000010) -3.00416613

FE_TOWARDZERO:
   exp(1.1)  = (0 10000000 10000000100010001000001) 3.00416589
  -exp(1.1)  = (1 10000000 10000000100010001000001) -3.00416589
```

which shows the floating-point number in decimal as well as the actual bits used to store the value. Note that Fig. 3.4 includes a comment line describing the actual command used to compile the program. The -frounding-math option is required otherwise fesetround() will not actually change the rounding mode.

If we look at the output we see that FE_TONEAREST, which is the default rounding mode, produces results which match what we expect since negating the value is not expected to alter the result other than the sign bit. We see this clearly by looking at the bits of the results; they are all the same except for the sign bit.

```
 1 #include <stdio.h>
 2 #include <fenv.h>
 3 #include <math.h>
 4 #include <string.h>
 5
 6 // compile: gcc rounding.c -o rounding -lm -frounding-math
 7
 8 int main() {
 9     fp_t fp;
10
11     printf("\nFE_TONEAREST:\n");
12     fesetround(FE_TONEAREST);
13     fp.f = exp(1.1);
14     printf("   exp(1.1) = (%s) %0.8f\n", ppx(fp.d), fp.f);
15     fp.f = -exp(1.1);
16     printf("  -exp(1.1) = (%s) %0.8f\n\n", ppx(fp.d), fp.f);
17
18     printf("FE_UPWARD:\n");
19     fesetround(FE_UPWARD);
20     fp.f = exp(1.1);
21     printf("   exp(1.1) = (%s) %0.8f\n", ppx(fp.d), fp.f);
22     fp.f = -exp(1.1);
23     printf("  -exp(1.1) = (%s) %0.8f\n\n", ppx(fp.d), fp.f);
24
25     printf("FE_DOWNWARD:\n");
26     fesetround(FE_DOWNWARD);
27     fp.f = exp(1.1);
28     printf("   exp(1.1) = (%s) %0.8f\n", ppx(fp.d), fp.f);
29     fp.f = -exp(1.1);
30     printf("  -exp(1.1) = (%s) %0.8f\n\n", ppx(fp.d), fp.f);
31
32     printf("FE_TOWARDZERO:\n");
33     fesetround(FE_TOWARDZERO);
34     fp.f = exp(1.1);
35     printf("   exp(1.1) = (%s) %0.8f\n", ppx(fp.d), fp.f);
36     fp.f = -exp(1.1);
37     printf("  -exp(1.1) = (%s) %0.8f\n\n", ppx(fp.d), fp.f);
38 }
```

Fig. 3.4 A program to display the effect of different rounding modes. Note that fp_t and ppx()
are defined in Fig. 3.2

The remaining directed modes act as expected. For FE_UPWARD we see that the
negative result is smaller than the positive, in magnitude, because it has been
rounded towards positive infinity. Similarly, we see the magnitudes reversed for
FE_DOWNWARD which is rounding towards negative infinity. Lastly, FE_TOWARDZERO
rounds towards zero giving the smaller magnitude result for both positive and negative cases. The floating-point bit patterns for FE_TONEAREST and FE_TOWARDZERO
are the same in each case except for the sign bit while the other two cases are similar
but the positive and negative bit patterns are flipped (for the significand).

If the default of "round to nearest" is so widely used, why are the other rounding modes present in the IEEE standard? One example of why can be found in [8] where directed rounding significantly improved the accuracy of the calculations. Regardless of specialized uses, in most cases it is completely unnecessary to modify the rounding mode from the default.

3.6 Comparing Floating-Point Numbers (IEEE 754)

The IEEE standard requires that floating-point numbers maintain an explicit ordering for comparisons. The simplest way to compare two floating-point numbers is to look at their bit patterns. For example, the following two numbers are as close as two 32-bit floating-point numbers can be without being equal. This means there are no representable floating point numbers between them. The numbers are,

```
0.500000000 = 0 01111110 00000000000000000000000 = 1056964608
0.500000060 = 0 01111110 00000000000000000000001 = 1056964609
```

where the floating-point value is on the left, the actual bit patterns are in the center and the equivalent 32-bit signed integer is on the right. Note that when viewed as an integer the smaller number, 0.5, is also smaller. This is no accident. The floating-point format is designed to make comparison of floating-point numbers as straightforward as comparing two signed integers where negative floating-point numbers are interpreted in two's-complement format. The only exception is zero since it can be signed or unsigned. But it is easy to check for this single case.

The C code for a floating-point comparison function that checks for the special case of signed zero is,

```
 1 typedef union {
 2     float f;
 3     int d;
 4 } sp_t;
 5
 6 int fp_compare(float a, float b) {
 7     sp_t x,y;
 8
 9     x.f=a;
10     y.f=b;
11
12     if ((x.d == (int)0x80000000) &&
13         (y.d == 0)) return 0;
14     if ((y.d == (int)0x80000000) &&
15         (x.d == 0)) return 0;
16
17     if (x.d == y.d) return 0;
18     if (x.d < y.d) return -1;
19     return 1;
20 }
```

where we introduce sp_t instead of fp_t which we used previously. The difference
is we now use a signed integer instead of an unsigned integer. The signed integer is
necessary so that the integer comparisons will treat negative floating-point numbers
properly.

The fp_compare function accepts two floats, a and b, as input and returns -1 if
$a < b$, 0 if $a = b$, and $+1$ if $a > b$. In lines 9 and 10 the input floats are placed into
two sp_t variables, x and y. This is so we can get the equivalent signed integer from
the floating-point bit pattern. Lines 12 and 13 are the special check for negative zero
compared to positive zero. In this case we want a return value of 0 since the standard
says these values are equal. The cryptic (int)0x80000000 is the bit pattern for
a negative zero number cast to a signed integer. If this is the first value and the
second is positive zero, consider them the same. Lines 14 and 15 make the same
comparison with the arguments reversed. Line 17 checks if the bit patterns are the
same and returns zero if they are to declare the values equal. Line 18 checks to see
if $a < b$ and returns -1 if they are. The only option left is that $a > b$ so we return
$+1$ in line 19. This completes the predicate function to compare two floating point
numbers using their integer bit patterns.

For completeness, let's define functions for the basic comparison operators. We
can then check these against the intrinsic operators in C for floating-point numbers.
The functions are,

```
1  int fp_eq(float a, float b) {
2       return fp_compare(a,b) == 0;
3  }
4
5  int fp_lt(float a, float b) {
6       return fp_compare(a,b) == -1;
7  }
8
9  int fp_gt(float a, float b) {
10      return fp_compare(a,b) == 1;
11 }
```

And we can use them with the following code,

```
1  int main() {
2       float a,b;
3
4       a = 3.13;   b = 3.13;
5       printf("3.13 == 3.13: %d %d\n", fp_eq(a,b), a==b);
6
7       a = 3.1;    b = 3.13;
8       printf("3.1 < 3.13: %d %d\n", fp_lt(a,b), a<b);
9
10      a = 3.2;    b = 3.13;
11      printf("3.2 > 3.13: %d %d\n", fp_gt(a,b), a>b);
12
13      a = -1;   b = 1;
14      printf("-1 < 1: %d %d\n", fp_lt(a,b), a<b);
15 }
```

which shows that comparing the integer representations gives the results we would expect,

```
3.13 = 3.13 : 1 1
3.10 < 3.13 : 1 1
3.20 > 3.13 : 1 1
-1.00 < 1.00 : 1 1
```

In this discussion we have ignored NaNs and infinity. NaNs, according to the standard, are unordered even when compared to themselves. Infinities of the same sign compare as equal while those of opposite signs should compare as expected. This is not the case with the simple example above in fp_compare which would need to be updated to compare infinities properly.

3.7 Basic Arithmetic (IEEE 754)

The standard mandates that conforming implementations supply the expected addition ($+$), subtraction ($-$), multiplication ($\times$) and division ($/$) operators. It also requires square root of a positive number and a fused multiply-add which implements $(xy) + z$ with no exceptions thrown during the multiplication. This combined operation only involves one rounding step, not the two that would happen if the multiplication was done separately from the addition.

Addition and Subtraction. The addition of two floating-point numbers is straightforward. Let's consider the steps necessary to calculate $8.76 + 1.34$. First, we represent each number in binary as:

```
8.76 → 8.760000229 = 0 10000010 00011000010100011110110
1.34 → 1.340000033 = 0 01111111 01010111000010100011111
```

where we immediately see that neither 1.34 nor 8.76 can be exactly represented using a 32-bit float and that we have again shown the binary representation separating the sign, exponent and significand. If we convert the bit pattern to actual binary we get,

$$8.76 \rightarrow 1.00011000010100011110110 \times 2^3$$
$$1.34 \rightarrow 1.01010111000010100011111 \times 2^0$$

where we see that the exponents are not the same. In order to add the significands we need to make the exponents the same first. So, we rewrite the smaller number so that it has the same exponent as the larger. This means we add the difference of the exponents to the exponent of the smaller number, $3 - 0 = 3 \rightarrow 3 + 0 = 3$, and shift the significand of the smaller number that many bit positions to the left to get,

$$8.76 \rightarrow 1.00011000010100011110110\,000 \times 2^3$$
$$1.34 \rightarrow 0.00101010111000010100011\,111 \times 2^3$$

with the bits that would have been shifted out of the 32-bit float separated from the rest of the significand by a space. Normally, the calculation is done with a higher precision than the storage format and rounding is applied. The top value is filled in by three zeros for the extended precision bits.

With the exponents the same the addition is simple, add the significand, bit by bit, just as one would add 2.5×10^3 and 4.6×10^3 to get 7.1×10^3,

$$
\begin{array}{r}
8.76 \rightarrow 1.000110000101000111101100 00 \times 2^3 \\
+\ 1.34 \rightarrow 0.00101010111000010100011\ 111 \times 2^3 \\
\hline
1.01000011001100110011001\ 111 \times 2^3
\end{array}
$$

If the answer does not have a single 1 to the left of the decimal point we shift the significand to the right until it does so that our answer is a normalized number. Naturally, if we shift the significand n bits to the right we need to raise the exponent by adding n to it. In the example above the answer is already normalized since the first, and only, bit to the left of the decimal point is 1.

There is one step remaining, which is to account for rounding. Assuming that the rounding mode is the default of "round to nearest" we need to consider the first 1 in the three lowest order bits of the answer, the three bits on the far right of the significand. In order to round up to the nearest floating-point number we add one to the right-most bit on the significand, before the space, to make $001 \rightarrow 010$. In this case, then, the final answer is,

$$
10.100000381 \rightarrow 1.0100001100110011001010 \times 2^3
$$

Subtraction works in the same way. Specifically, the steps to add or subtract two floating-point numbers x and y are,

1. Raise the exponent of the number with the smaller exponent by the difference between the larger and smaller exponent. Also, shift the significand of the smaller number to the right by the same number of bits. This causes both numbers to have the same exponent. If the smaller number was y the shifted number is y'.
2. Add or subtract the significands of x and y' as they now have the same exponent. Put the answer in z.
3. If there is not a single 1 to the left of the decimal point in z, shift the significand of z to the right until there is. Increase the exponent of z by the number of bits shifted.

Multiplication. If we want to multiply two numbers in scientific notation we multiply the significands and add the exponents,

$$
(a \times 10^x)(b \times 10^y) = ab \times 10^{x+y}
$$

similarly, for two floating-point numbers expressed in binary, we multiply the significands and add exponents then normalize. So, to multiply 3.141592 by 2.1, we have,

```
3.141592 → 3.141592026 → 0  10000000  10010010000111111011000
2.1        → 2.099999905 → 0  10000000  00001100110011001100110
```

where the first column is the real number we want to multiply, the second column is the actual floating-point number we get when using *binary32* and the last column is the actual bit pattern in memory. Using the bit pattern we can write these values in binary as,

$$3.141592 \rightarrow 1.10010010000111111011000 \times 2^1$$
$$2.1 \quad \rightarrow 1.00001100110011001100110 \times 2^1$$

which means that our product will have an exponent of $1 + 1 = 2$. We can multiply the significands, bit by bit, which is tedious on paper so instead we approximate the product using only seven bits after the radix point,

```
              1.1001001
    ×         1.0000110
    ─────────────────────
                      0
              1110010010
             1110100100
                   . . .
        110010010000000
    + 1.10100100110110
```

where we have already moved the radix point to the proper position in the final sum. The . . . represents a series of all zero places corresponding to the string of four zeros in the multiplier. The last line shows the last partial product. The full product of the significands is `1.10100110001110101101111`.

Since the significand product already has a leading value of one we do not need to shift in order to get a leading bit of one. For example, if the significand product started with `10.00101...` we would shift the radix point one position to the left so make the significand `1.000101...` and increase the exponent by one position.

So, finally, our now approximate product (because we dropped many significand bits) can be represented as,

$$1.10100100110110 \times 2^2 \rightarrow 6.5756835938$$

while the full precision product is $3.141592 \times 2.1 = 6.5973429545$.

In summary, then, the steps to multiply two floating-point numbers x and y are,

1. Add the exponents of x and y.
2. Multiply the significands of x and y.
3. Shift the product of the significands to get a single leading one bit to the left of the radix point. Adjust the exponent as well by adding one for every bit position the product of the significands is shifted to the right. This is equivalent to shifting the radix point to the left.
4. Reconcile the signs so that final product has the expected sign.

This concludes our look at floating-point arithmetic. Next we consider an important part of the IEEE 754 standard we have neglected until now, exceptions.

3.8 Handling Exceptions (IEEE 754)

Not all floating-point operations produce meaningful results for all operands. An important part of the IEEE 754 standard addresses what to do when exceptions to normal floating-point processing arise. The standard recommends a default set of actions for these cases and then allows implementations to give users the power to trap and respond to exceptions. In this section we learn what these exceptions are, the default responses the standard recommends for them, and how to trap exceptions to bring them under user control in C. It is important to remember that floating-point exception handling is distinct from the concept of exceptions in programming languages, though they are, of course, related and could be intermingled. Modern processors, including the x86 line, process floating-point exceptions in hardware and as such are programming language independent.

Exceptions and Default Handling The standard defines five types of exceptions: *inexact, invalid, underflow, overflow* and *divide-by-zero*:

Exception Type	Description	Examples
inexact	When the result is not exact due to rounding.	$2.0/3.0$
invalid	When the operands are not valid for the operation.	$0/0, 0 \times \infty, \sqrt{x}, x < 0$
underflow	When the result is too small to be accurate or when subnormal.	$(1.17549435 \times 10^{-38})/3.0$
overflow	When the result is too large to represent	$(3.40282347 \times 10^{38})^2$
divide-by-zero	When a finite number is divided by zero	$1/0$

The standard also defines default behavior for these exceptions. The goal of the default behavior is to allow the computation to proceed by substituting a particular value for the exception or to just ignore the exception altogether. For example, the default behavior for an *inexact* exception is to ignore it and continue with the calculation. The default behavior for an *invalid* exception is to generate a NaN and then propagate the NaN through the remainder of the calculation. If *divide-by-zero* happens the result is passed on as $\pm\infty$ depending upon the signs of the operands.

The *overflow* exception is raised when the result does not fit. What the default behavior is depends upon the rounding mode in effect at the time as well as the sign of the number. Specifically,

Round to Nearest	Return $\pm\infty$ depending upon the sign of the result.
Round Towards Zero	Return the largest number for the format with the sign of the result.
Round Towards $+\infty$	Return $+\infty$ if positive, largest negative number if negative.
Round Towards $-\infty$	Return $-\infty$ if negative, largest positive number if positive.

Lastly, *underflow* is raised when the number is too small. The implementer is allowed to determine what "too small" means. The default result returned will be the subnormal number or zero.

Trapping Exceptions For many programs the default behavior is perfectly acceptable. However, for total control over floating-point operations the standard allows for traps which run user code when an exception happens. For x86 architectures, traps are in hardware but can be used from programming languages such as C. When a trap executes, or is used by a higher-level library, the programmer is able to control what happens when exceptions occur. Let's take a look at how one can do this using gcc.

The first thing to be aware of is that by default, the gcc compiler is not 100 % compliant with IEEE 754. So, to get the performance we expect we need to be sure to use -frounding-math and -fsignaling-nans when we compile. Also, we need to include the fenv.h library. With these caveats, the set of functions we can use depends upon the way in which we want to control floating-point exceptions. We can throw a signal when they happen (SIGFPE) or we can test individual operations for individual exceptions. Let's look at using a signal first.

A signal is a Unix concept, recall we assume a 32-bit Linux implementation, which is a message to a program to indicate that a particular event has happened. There are many possible signals, but the one we want is SIGFPE to signal a floating-point error. We need to enable floating-point exceptions, then arrange a signal handler like so,

```
 1 #include <stdlib.h>
 2 #include <stdio.h>
 3 #include <signal.h>
 4 #include <fenv.h>
 5
 6 void fpe_handler(int sig) {
 7     if (sig != SIGFPE) return;
 8     printf("Floating point exception!\n");
 9     exit(1);
10 }
11
12 int main() {
13     float a=1, b=0, c;
14
15     feenableexcept(FE_INVALID   | FE_DIVBYZERO |
16                    FE_OVERFLOW | FE_UNDERFLOW);
17     signal(SIGFPE,  fpe_handler);
18
19     c = a/b;
20 }
```

where lines 1 through 4 include necessary libraries. Lines 6 through 10 define our signal handler. We check to see if the signal is actually a floating-point exception (line 7) and then report it and exit. In main we define some floats and then enable floating-point exception traps by calling `feenableexcept()` with the bit masks for the floating-point exceptions we want to trap. Line 17 ties the `SIGFPE` signal to our handler, `fpe_handler()`. Finally, line 19 tries to divide by zero which will trigger the exception. This code should be compiled with,

$$\$gcc\ fpe.c - ofpe - lm - frounding - math - fsignaling - nans$$

assuming `fpe.c` to be the name of the file.

Another way to work with floating-point exceptions is to explicitly check if any were raised during a calculation. In this case we need the `feclearexcept()` and `fetestexcept()` functions to clear any exceptions prior to the calculation and then to test for particular exceptions after. For example, this program,

```
 1  #include <stdio.h>
 2  #include <signal.h>
 3  #include <fenv.h>
 4
 5  int main() {
 6      float c;
 7
 8      feclearexcept(FE_ALL_EXCEPT);
 9      c = 0.0/0.0;
10      if (fetestexcept(FE_INVALID) != 0)
11          printf("FE_INVALID happened\n");
12
13      feclearexcept(FE_ALL_EXCEPT);
14      c = 2.0/3.0;
15      if (fetestexcept(FE_INEXACT) != 0)
16          printf("FE_INEXACT happened\n");
17
18      feclearexcept(FE_ALL_EXCEPT);
19      c = 1.17549435e-38 / 3.0;
20      if (fetestexcept(FE_UNDERFLOW) != 0)
21          printf("FE_UNDERFLOW happened\n");
22
23      feclearexcept(FE_ALL_EXCEPT);
24      c = 3.40282347e+38 * 3.40282347e+38;
25      if (fetestexcept(FE_OVERFLOW) != 0)
26          printf("FE_OVERFLOW happened\n");
27
28      feclearexcept(FE_ALL_EXCEPT);
29      c = 1.0 / 0.0;
30      if (fetestexcept(FE_DIVBYZERO) != 0)
31          printf("FE_DIVBYZERO happened\n");
32  }
```

will produce this output,

```
FE_INVALID happened
FE_INEXACT happened
FE_UNDERFLOW happened
FE_OVERFLOW happened
FE_DIVBYZERO happened
```

when compiled with,

```
$ gcc fpe_traps.c -o fpe_traps -lm -frounding-math -fsignaling-nans
```

thereby showing how to explicitly test for floating-point exceptions. Note, just as `feenableexcept()` took an argument that was the bitwise OR of one or more exceptions, so too will `fetestexcept()` so that in order to test for *invalid* and *divide-by-zero* one would call,

```
int i = fetestexcept(FE_INVALID | FE_DIVBYZERO)
```

and then check for either with `i != 0` or one explicitly with `(i & FE_INVALID) != 0`.

3.9 Floating-Point Hardware (IEEE 754)

For this section we assume a system using modern x86 processors. These processors implement an 80-bit extended precision version of floating-point in hardware. This is why floating-point calculations are so fast when in the past floating-point was generally, especially for smaller systems, done in software. This is still true in the large embedded processor world which we will of necessity neglect entirely.

An x86 extended precision float is stored in registers as,

where it is very important to note that unlike the IEEE storage formats, there is no implied leading 1 bit for the significand. Instead, it is given explicitly as the first bit marked a "I" above. The sign (S) is 0 for positive and 1 for negative. The exponent (E) is stored as excess-16383 and the fractional part of the significand (M) is 63-bits long. This means that a floating-point number (n) is,

$$n = (-1)^s i_0.d_{62}d_{61}\ldots d_0 \times 2^{e-16383}$$

where there is now an explicit integer part (i_0) and the binary significand ($d_{62}\ldots d_0$). How many bits of the significand are actually used in a particular calculation is a function of the processor (32-bit vs 64-bit), the compiler, and various compiler settings. We ignore these important but intricate details here and instead focus on the representation.

Just as an IEEE *binary32* or *binary64* float uses certain bit combinations to indicate NaN and infinity so does the hardware version. The explicit integer part helps in quickly determining the type of number that is being worked with. Table 3.1 illustrates how to decide whether or not the float is normal, subnormal, zero, infinity, NaN or invalid.

(a) Normal, Subnormal and Zero:

Exponent	I	M	Value
000000000000000	0	all 0	zero (sign ignored)
000000000000000	0	non-0	subnormal
any but all 1's	1	any	normal

(b) Infinity, Not-a-Number, and Invalid (exponent = 111111111111111):

I	M_{62}	$M_{61}\ldots M_0$	Value
1	0	0	$\pm\infty$ (uses sign)
1	0	non-0	signaling NaN
1	1	0	invalid result (e.g., $0/0$)
1	1	non-0	quiet NaN

Table 3.1 How to interpret the bit patterns of an x86 extended precision floating-point number. Only the meaningful combinations of bits are shown, all others should be interpreted as *invalid*. When the CPU performs floating-point calculations it uses this format converting to and from IEEE 754 formats like *binary32* and *binary64* as needed. In *(a)* the exponent and integer part are used to decide between zero, subnormal numbers and normal numbers. All normal numbers have an integer part of 1 matching the implied leading 1 bit of the IEEE formats. For *(b)* the exponent is the special case of all ones. The integer part and leading bit of the significand determine the number type

The top part of the table illustrates normal, subnormal and zero. As is the case for IEEE, there are two possible representations for zero, signed and unsigned. The sign is ignored in calculations. An integer part of zero indicates a subnormal number. If the integer part is 1 and the exponent is any value except all zeros or all ones the number is a normalized floating-point number. In the bottom part of the table are the exceptional values. All of these have the integer part set to 1 and the exponent set to all ones. These values are distinguished based on the value of the leading fractional part of the significand and the remaining bits of the significand. If the leading bit, M_{62}, is zero the number is infinity (all remaining significand bits zero) or a signaling NaN. Recall that a signaling NaN raises an exception. If M_{62} is set the number is either a quiet NaN (remainder of the significand bits are not zero) or an invalid number (all zero bits).

The bit patterns covered in Table 3.1 are not exhaustive. There are other combinations which could be present. In the past some of these were used but now all are considered invalid and will not be generated by the floating-point hardware.

The large number of bits in the significand means that 32-bit and 64-bit floating point operations can be performed with full precision and proper rounding of results. Above we discussed the basic arithmetic operations called out by the IEEE 754 standard. There is one we intentionally ignored at the time. This is the fused multiply-add operation, $xy + z$, which first calculates xy and then adds z *without* rounding after the multiplication. If this were not done the expression would involve two roundings, one for the multiplication and another for the addition. This is unnecessary and with the extended precision of the x86 floating-point format will result in fewer rounding artifacts. The $xy + z$ form was chosen because it appears frequently in expressions and a smart compiler would look for these expressions so as to use the fused multiply-add capability whenever possible.

3.10 The Elephant in the Living Room (Pitfalls of Floating Point)

Programmers use floating-point numbers very frequently and sometimes without careful consideration of the consequences from the imprecision which may arise. In this section we will look at several examples of where floating-point numbers have led to disaster, often with the loss of human life. After these cautionary tales we will offer some general (and hopefully helpful) advice on how to improve our chances of creating a correctly functioning program involving floating-point values.

Patriot Missile Failure During the Gulf War (August 1990 through February 1991) the Iraqi army employed Soviet-built Scud missiles. The United States responded by installing Patriot anti-missile batteries to shoot down approaching Scuds. On February 25, 1991 a Scud missile hit a US army barracks in Dharan, Saudi Arabia, killing 28 solders. The base was protected by Patriot missile batteries yet the battery in range of the Scud failed to fire. The reason why is due to round-off error in floating-point calculations. The full report of the incident is available from the United States General Accounting Office (GAO) [9] but the essence of the problem has to do with trying to represent a simple value, in this case 0.1, in a floating-point format which was itself then truncated to fit into a fixed-point number.

The Patriot missile software tracked the position of incoming Scuds using radar. It tracked time by counting tenths of a second and used a 24-bit fixed-point representation of the floating-point value for 0.1 to convert to seconds. The only problem is that in binary the value of 0.1 is an infinitely repeating decimal,

$$0.1 = 0.0001100110011001100110011001100\ldots_2$$

that simply truncating at 24-bits gives, $0.00011001100110011001100_2$ meaning an error of $0.0000000000000000000000011001100110011001100\ldots_2 \approx 0.000000095$ seconds was in introduced for each truncation. This is not much, but if the missile battery was left in operation for a long period of time the error would accumulate.

The GAO report indicates that Israeli forces, also operating Patriot batteries, noticed that after approximately eight hours of operation that the targeting was off by about 20 %. A software fix was created and scheduled to be sent to the batteries. However, no long running tests were performed to see how long the battery could be in operation before the timing error was catastrophic. Battery operators were warned that long runtimes might cause inaccuracy in targeting but were not given any notion of what "long runtime" might mean. On February 25, 1991 the Patriot battery in Dharan had been in continuous operation for about 100 hours. This means the small timing error introduced by the truncation of 0.1 to 24-bits was now multiplied by the number of tenths of a second in 100 hours or 3,600,000 causing a discrepancy of about 0.34 seconds. During this time a Scud missile would cover about half of a kilometer putting it out of range of the battery. Since the battery thought the missile was out of range it did not fire and the missile went on to kill 28 soldiers. The software patch arrived at Dharan on February 26th.

Ariane 5 Explosion The European Space Agency has a successful track record when it comes to launching sophisticated spacecraft. However, the maiden flight of its Ariane 5 rocket on June 4, 1996, was a total loss, to the tune of $500 million USD, all because one of the software modules attempted to put a 64-bit floating-point value into a 16-bit signed integer.

Just under 40 seconds after its launch from Kourou, French Guiana, the rocket self-destructed. The software putting a 64-bit floating-point value into the signed 16-bit integer had caused an overflow (i.e., the value exceeded 32,767) which caused the rocket to believe it was off-course. It therefore fired thrusters to "correct" its course which put so much stress on the rocket that its self-destruct was triggered. The final report of the inquiry board summarizes the events [10].

An Experiment The examples above show that floating-point can sometimes lead to fatal and costly disasters. More practically, floating-point, in combination with compilers, hardware, and even the form of the equations used, can lead to wildly inaccurate results. Consider the following experiment inspired by previous work by Colonna [11].

A favorite equation used to illustrate the onset of chaotic behavior is the logistic map, $x_{i+1} = rx_i(1 - x_i)$ with r a constant between 1 and < 4 and x a value from $(0, 1)$. This map exhibits very interesting behavior as an initial x_0 is iterated repeatedly and as r varies from say 2 up to just under 4. If one were to select a starting value of 0.1 (or any other) and run the iteration forward a thousand times and then plot the next thousand as a function of r one would get Fig. 3.5 which is known as a bifurcation plot. This plot shows the period-doubling route to chaotic behavior as r increases. There is a deep correlation between the structure of this plot and the famous Mandelbrot set. Of interest to us is the sequence of values one gets if r is in the chaotic regime for different but algebraically equivalent forms of the logistic equation.

We can rewrite the logistic equation as,

$$x_{i+1} = rx_i(1 - x_i) = rx_i - rx_i x_i = x_i(r - rx_i) = rx_i - rx_i^2$$

where each of the four forms will be computed in our test program. The program itself is straightforward,

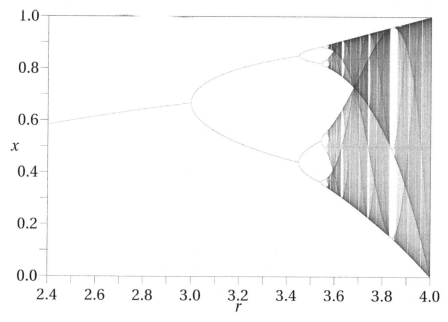

Fig. 3.5 The bifurcation plot of the logistic map. The plot shows several hundred x values from the iterated logistic map for each r value as a function of r. This map illustrates the period-doubling route to chaotic behavior. For the experiment $r = 3.8$ which is in the chaotic region ("Logistic Map" by Jordan Pierce, Creative Commons CC0 1.0 Universal Public Domain Dedication License)

```
 1  #include <stdio.h>
 2  #include <math.h>
 3
 4  int main() {
 5      double x0,x1,x2,x3;
 6      double r = 3.8;
 7      int i;
 8
 9      x0 = x1 = x2 = x3 = 0.25;
10
11      for(i=0; i < 100000; i++) {
12          x0 = r*x0*(1.0 - x0);
13          x1 = r*x1 - r*x1*x1;
14          x2 = x2*(r - r*x2);
15          x3 = r*x3 - r*pow(x3,2);
16      }
17
18      printf("      x0            x1            x2            x3\n");
19
20      for(i=0; i < 8; i++) {
21          x0 = r*x0*(1.0 - x0);
22          x1 = r*x1 - r*x1*x1;
23          x2 = x2*(r - r*x2);
24          x3 = r*x3 - r*pow(x3,2);
25          printf("%0.8f  %0.8f  %0.8f  %0.8f\n", x0,x1,x2,x3);
26      }
27  }
```

This program iterates each of the four versions of the logistic map and outputs the final eight iterates after an initial burn in of 100,000 iterations. Note that the program uses double precision values and that the initial value is 0.25 which is not a repeating binary. The program produces this as output,

x0	x1	x2	x3
0.76670670	0.65123778	0.80925715	0.38542453
0.67969664	0.86308311	0.58656806	0.90011535
0.82729465	0.44904848	0.92152269	0.34164928
0.54293721	0.94013498	0.27481077	0.85471519
0.94299431	0.21386855	0.75730128	0.47187310
0.20427297	0.63888941	0.69842500	0.94699374
0.61767300	0.87669699	0.80038457	0.19074708
0.89738165	0.41077765	0.60712262	0.58657800

when compiled with,

```
$ gcc logistic.c -o logistic -lm
```

where we already see the effect of the different way the compiler has decided to evaluate the logistic expressions even though algebraically they are identical. If we compile again with optimization,

```
$ gcc logistic.c -o logistic -lm -O2
```

we get,

x0	x1	x2	x3
0.18402020	0.67978266	0.32056222	0.92875529
0.57059572	0.82717714	0.82764792	0.25144182
0.93106173	0.54322944	0.54205800	0.71523156
0.24390598	0.94289862	0.94327827	0.77396643
0.70078024	0.20459509	0.20331661	0.66478109
0.79681171	0.61839658	0.61552007	0.84681933
0.61523066	0.89673255	0.89928943	0.49292215
0.89954320	0.35189247	0.34415822	0.94980964

which is different still. Clearly, the way in which an expression is coded matters significantly. Which value, if any, is most correct? How can one tell? One way would be to use arbitrary precision floating-point, which we do not have the space in this book to discuss (see MPFR [12]). Alternatively, one could use rational arithmetic and convert back to floating-point at the end. For an explanation of rational arithmetic see Chap. 4 but be warned that the iteration will quickly lead to fractions with very large numerators and denominators and require a great deal of memory.

Another Experiment Let's consider another experiment. In this case we want to compute e^x for any x using the Taylor series expansion,

$$e^x = 1 + x + \frac{x^2}{2!} + \frac{x^3}{3!} + \cdots = 1 + \sum_{i=1}^{\infty} \frac{x^i}{i!}$$

which we translate into Python as,

```
1  def fact(n):
2      if (n == 1):
3          return 1.0
4      return n * fact(n-1)
5
6  def exponential(x, tol=1e-10):
7      ans = 0.0
8      term= 1.0
9      p = 1.0
10
11      while (abs(term) > tol):
12          ans += term
13          term = x**p / fact(p)
14          p += 1
15
16      return ans
```

where we accumulate terms of the series until one of the terms is smaller than a given threshold. If we run this program for positive values of x and compare them with the output of a gold standard exp(x) function (recall that Python floats are double precision), we get,

```
20   4.8516519540979016e+08   4.8516519540979028e+08
21   1.3188157344832141e+09   1.3188157344832146e+09
22   3.5849128461315928e+09   3.5849128461315918e+09
23   9.7448034462489052e+09   9.7448034462489033e+09
24   2.6489122129843472e+10   2.6489122129843472e+10
25   7.2004899337385880e+10   7.2004899337385880e+10
```

where the first column is the argument, the second is the output of exponential() and the third is the output of our gold standard function. Clearly, our Taylor series approximation for e^x is working quite well.

Now, let's run again but this time make $x < 0$. In this case we get,

```
-20   4.9926392547479328e-09   2.0611536224385579e-09
-21   2.7819487708032662e-08   7.5825604279119066e-10
-22  -2.4369901527907362e-08   2.7894680928689246e-10
-23   3.1518333727167153e-09   1.0261879631701890e-10
-24   3.7814382919759864e-07   3.7751345442790977e-11
-25   1.1662837734562158e-06   1.3887943864964021e-11
```

Not so good. What happened and how can we fix it? What happened is that the Taylor series for e^x is made up of terms that are alternately even and odd powers of x. This means that the sign of the term changes from term to term. In other words, the series sum is made up of values that are wildly different from each other. Consider the approximation for e^{-20},

$$e^{-20} = 1 - 20 + \frac{20^2}{2} - \frac{20^3}{6} + \frac{20^4}{24} - \frac{20^5}{120} + \frac{20^6}{720} - \cdots$$
$$= 1 - 20 + 200 - 1333.\overline{3} + 6666.7 - 26666.7 + 88888.9 - \cdots$$

where the effect of each term changing sign is that the sum to term n is always the opposite sign as term $n + 1$. This addition of two very different numbers leads to a rapid loss of precision making the approximation functionally useless even though mathematically it is valid.

The fix in this case is to use only positive x and to recall that $e^{-x} = 1/e^x$. If we run our Python program again with only positive x and report 1.0/exponential(x) we greatly improve our final results,

```
-20   2.0611536224385583e-09   2.0611536224385579e-09
-21   7.5825604279119097e-10   7.5825604279119066e-10
-22   2.7894680928689241e-10   2.7894680928689246e-10
-23   1.0261879631701888e-10   1.0261879631701890e-10
-24   3.7751345442790977e-11   3.7751345442790977e-11
-25   1.3887943864964019e-11   1.3887943864964021e-11
```

thereby proving that thought is always required when expecting meaningful results from floating-point operations.

Rules of Thumb We end this chapter with some "rules of thumb" regarding the use of floating-point numbers. These rules are not absolute, of course, and others could be given, but they will hopefully be found helpful and cause you to pause and consider carefully when writing floating-point programs.

1. Do not use floating-point numbers if integers will suffice. As a corollary, do not use floating-point numbers if exact computation is required, rather, scale if a fixed number of decimals will always be used. For example, if the values are dollars do the computations in cents and divide by 100 when done.
2. Do not represent floating-point numbers in text form in files or as strings unless absolutely necessary. If necessary, use scientific notation and be sure to have enough digits after the decimal point of the mantissa to represent the significand accurately. For a 32-bit float use at least 8 digits. For a 64-bit float use at least 16. For example, in Python,
   ```
   >>> from math import pi
   >>> float("%0.15g" % pi) == pi
   False
   >>> float("%0.16g" % pi) == pi
   True
   ```
   where the 0.15g format specifier uses 15 digits after the decimal which is not enough, 16 are required so that the text representation of π, when converted back to a floating-point number, matches the original. Floating-point numbers are often in XML files and frequently without enough precision.
3. The spacing between floating-point numbers increases as the number increases. If working with large numbers, rescale, if possible, to map to values closer to one.
4. Comparing two floating-point values for equality can be problematic if they were computed using a different code sequence. If possible, replace expressions like a == b with abs(a-b) < e where e is a tolerance value. If the numbers are "close enough" so that their difference is less than e we can regard them as equal (or, not equal). Even this is dangerous in that the choice of e depends upon the scale of the values being worked with. Absolute comparison for equality

involves treating the floating-point numbers as integers, as we did above using the C union, and comparing these integers bit by bit. IEEE floating-point numbers are ordered so two floats are only truly equal if their bit patterns are identical.

5. Subtracting two nearly equal numbers results in a large loss of precision as the result is made up of only the lowest order bits of the significand. Generally, this should be avoided. Naturally, the rule above does just this which is why it is best to compare integer representations directly when possible and absolute precision is needed.

6. For real numbers we know that $(ab)c = a(bc)$ however this is sometimes not the case for floating-point numbers due to rounding. Use care when assuming associativity.

The rules above are ad hoc examples of a large research area known as *numerical stability* or *numerical analysis*. An important part of numerical stability involves an analysis of the equations being implemented by the computer and a reworking of those equations to minimize calculations requiring floating-point operations known to lead to poor results. The stability of numerical algorithms is a vast field as might be expected. For a scholarly introduction, especially its early chapters, see Higham [13]. For a more popular treatment see Acton [14].

3.11 Chapter Summary

In this chapter we took a thorough look at floating-point numbers and how they are represented in a computer. We discussed just what we mean by "floating-point" and talked about some of the history surrounding floating-point and computers. We compared two popular (though one is largely disused now) floating-point representations. We then explored the main parts of the IEEE 754 standard covering representations, rounding modes, comparisons, basic arithmetic, and exceptions. Next, we discussed the relationship between the standard and common floating-point hardware. Lastly, we covered some of the many pitfalls associated with floating-point numbers by examining several catastrophic failures due to floating-point math.

Exercises

3.1. Convert the following 32-bit IEEE bit patterns to their corresponding floating-point values:

(a) 0 10000000 10010010000011111011011
(b) 0 10000000 01011011111100001010100
(c) 0 01111111 10011110001101110111101
(d) 0 01111110 01100010111001000011000
(e) 0 01111111 01101010000010011110011

Where the first number is the sign, the second group of 8 bits is the exponent and the remaining bits are the significand. It will help to write a computer program to handle the powers of two in the significand.

3.2. Assume a simple floating-point representation that uses four bits in the significand, three bits for the exponent, and one for the sign. The exponent is stored in excess-3 format and all exponent values are valid. In this case we have the following representations,

```
 4.25 = 0 101 0100
  1.5 = 0 011 1000
 -1.5 = 1 011 1000
-4.25 = 1 101 0100
```

Work out the following sums and express the answer in bits. Use a floating-point intermediate with eight bits for the significand and then round your answer to the nearest expressible value. Find `4.25 + 1.5` and `-4.25 + -1.5`. **

3.3. Modify `fp_compare` to handle infinities properly.

3.4. Summing the values of an array is a common operation. If the naive implementation is used, simply adding each new array element to the accumulated sum of all the previous, there will be an overall round-off error proportional to the number of terms added. For a small array this is negligible but for a large array it may matter. One way to avoid the accumulated round-off error is to use *pairwise summation* [13]. This is a recursive divide-and-conquer algorithm which checks to see if the input array length is below a cutoff value, say five elements, and if so it returns their sum by simply adding them together. Otherwise, it takes the length of the input array, divides it by two using integer division, and calls itself on each of those subarrays adding the return values together to form the final output value. In pseudo-code,

```
function pairwise(x) {
    if (length(x) <= 5) {
        s = x[0] + x[1] + ...
    } else {
        n = length(x) / 2
        s = pairwise(x[0..n-1]) + pairwise(x[n..])
    }
    return s
}
```

Implement this function in C and Python. Demonstrate the effect of round-off by comparing the output of this function with the naive summation for input arrays of 10,000 elements or more. Use a pseudo-random number generator to generate values in the range $[0, 1)$. **

References

1. Muller, J., Brisebarre, N., *et al.* Handbook of Floating-Point Arithmetic, Birkhäuser Boston (2010).
2. Saracino D., Abstract Algerbra: A First Course, Addison-Wesley (1980).
3. Goldberg, D., "What every computer scientist should know about floating-point arithmetic." ACM Computing Surveys (CSUR) 23.1 (1991): 5–48.
4. Randell, B., The Origins of Digital Computers-Selected Papers. Springer-Verlag, (1982).
5. Rojas, R., "Konrad Zuse's legacy: the architecture of the Z1 and Z3." Annals of the History of Computing, IEEE 19.2 (1997): 5–16.
6. Cody, W., Waite, W. Software Manual for the Elementary Functions, Prentice-Hall, (1980).
7. Monniaux, D., "The pitfalls of verifying floating-point computations." ACM Transactions on Programming Languages and Systems (TOPLAS) 30.3 (2008): 12.
8. Rump, SM., "Accurate solution of dense linear systems, Part II: Algorithms using directed rounding." Journal of Computational and Applied Mathematics 242 (2013): 185–212.
9. B-247094, Report to the House of Representatives. Washington, D.C.: GAO, Information Management and Technology Division, 4 Feb. 1992, url: www.fas.org/spp/starwars/gao/im92026.htm (retrieved 15 Oct 2014).
10. Lions, JL (chair), ARIANE 5 Flight 501 Failure Report by the Inquiry Board, Paris, 19 July 1996.
11. Colonna, JF, The Subjectivity of Computers, http://www.lactamme.polytechnique.fr/descripteurs/subject.01..html. (retrieved 15 Oct 2014).
12. http://www.mpfr.org/. (retrieved 15 Oct 2014).
13. Higham, N. Accuracy and stability of numerical algorithms, Siam, (2002).
14. Acton, F., Real computing made real: Preventing Errors in Scientific and Engineering calculations, Courier Dover Publications, (2013).

Part II
Other Representations

Chapter 4
Big Integers and Rational Arithmetic

Abstract Big integers differ from standard integers in that they are of arbitrary size; the number of digits used is limited only by the memory available. In this chapter we look at how big integers are represented in memory and how to perform arithmetic with them. We also discuss some implementations which might be of use when using programming languages that do not support big integers natively. Next we examine rational arithmetic with big integers. Finally, we conclude with some advice on when it might be advantageous to use big integers and rational numbers.

4.1 What is a Big Integer?

Big integers, also known as bignums, bigints, multiple precision, arbitrary precision, or infinite precision integers, are integers that are not limited to a fixed number of bits. The standard representation for integers in computers involves a fixed number of bits, typically 8, 16, 32, or 64. These numbers therefore have a fixed range. A big integer, however, uses as much memory as is necessary to represent the digits and therefore is only limited in range by the memory available to it.

Several programming languages include big integers as part of their specification including Python. In Python we have an interpreter that seamlessly moves between fixed-width integers and big integers, of data type `long`, depending upon the result of a calculation. The transition point depends upon the version of the Python interpreter in use. The easy way to check this limit is to import the `sys` library and get the value for `sys.maxint`. This value is the largest positive integer which the interpreter will store in the standard representation. If this value is exceeded, the interpreter will switch to a big integer representation. For example, on a 32-bit system we have the following,

```
>>> import sys
>>> sys.maxint
    2147483647
>>> sys.maxint + 1
```

© Springer International Publishing Switzerland 2015

R.T. Kneusel, *Numbers and Computers*, DOI 10.1007/978-3-319-17260-6_4

```
        2147483648L
>>> -sys.maxint
        -2147483647
>>> -sys.maxint - 1
        -2147483648
>>> -sys.maxint - 2
        -2147483649L
```

where the Python prompt is given as >>> and replies from the interpreter are shown
without the prompt and indented for clarity.

Several important things to note are happening in this example. First, we import
the sys library and ask for the value of maxint to which Python replies it is
$2,147,483,647$ which we recognize as the largest positive two's complement
32-bit integer. This is to be expected on a 32-bit system. If we take this value and
add one, which results in a value that cannot be represented by a 32-bit two's complement integer, we get the answer we expect with an extra "L" after it. This extra
letter is Python's way of saying the data type is now long, meaning a big integer.
If we negate the maxint value we get a two's complement integer. If we then subtract one from this value we still get a two's complement integer, which is expected
because a two's complement integer gives us one extra negative number compared
to one's complement notation. However, if we subtract two we get our expected
answer but this time Python has moved to a big integer representation.

In this chapter we will look at ways in which big integers are represented in memory and some algorithms for dealing with big integer input and output when the base
is a convenient one. We then look at arithmetic and explore some of the more common big integer algorithms. Next, we look at some more sophisticated big integer
algorithms which are suitable when the numbers involved become quite large. We
then look at rational arithmetic, a natural place to use big integers. We follow this
with a, by necessity, time-sensitive view of key big integer implementations covering libraries for languages without big integers which we expect to be long-lived and
certain languages which implement big integers natively. We conclude with some
comments on when one might wish to resort to big integers or rational arithmetic.

4.2 Representing Big Integers

We know that place notation represents numbers using a given base, B, and the coefficients of the powers of that base which sum to the number we want to represent.
Specifically, we can represent a number in any base with,

$$abcd.efg_B = a \times B^3 + b \times B^2 + c \times B^1 + d \times B^0 + e \times B^{-1} + f \times B^{-2} + g \times B^{-3}$$

where we typically use $B = 10$ as the base. When we represent a big integer in
memory we would like to use a base which allows us to accomplish two things
simultaneously,

1. Use a base which makes conversion between text input and output straightforward.
2. Use a base in which products of the "digits" of our base fit in a standard word.

If we have a 32-bit system and look at using signed 32-bit integers as the way to represent digits we need a base where any two digits, when multiplied, give us a value that fits in a 32-bit integer. We will consider representation of signs for big integers later. This will satisfy the second requirement above. What about the first? If one looks at the common number bases used in computer systems we see that they are all powers of two. Octal is base $2^3 = 8$ while hexadecimal is base $2^4 = 16$. In these cases we saw that it was particularly simple to change a binary number into an octal or hexadecimal number and vice versa. This is clue we can use when choosing a base for our big integers. Since we would like to make things convenient for users we want to be able to input and output base ten numbers. This suggests using a power of ten for our big integer base, but which power? Since in this book we are assuming a 32-bit computer we want a base such that the base squared fits in a signed 32-bit integer. If we choose $B = 10,000$ we see that $B^2 = 100,000,000$ which does fit nicely in a signed 32-bit integer. However, if we go to the next power of ten, $B = 100,000$ we see that $B^2 = 10,000,000,000$ which is too large to fit in 32-bits. So, let's work with $B = 10,000$. Why choose such a large base? Why not simply use $B = 10$? The answer is that the larger the base the smaller the number of unique digits which are required to represent a large number. In binary we need eight bits to represent a number which in base ten is only three digits long and is only two digits in hexadecimal. So, the larger the base, the less memory we will need to store our numbers. As we are, by definition, working with big numbers we would like to be as economical as possible. If we were willing to work harder to handle input and output we could, naturally, use an even larger base. The largest value that fits in an unsigned 32-bit integer is $4,294,967,295$ and we see that $65,535^2 = 4,294,836,225$ will fit meaning that is the largest base we could use. However, as it is not a power of ten, we would have to work harder in input and output and since the larger base does not change the algorithms we will be exploring in this chapter we are justified in working with the smaller but handier base of $10,000$. An additional nice feature of our choice of base is that it fits in a *signed* 32-bit integer allowing us to make our internal representation quite compact as we will see.

We have selected to represent our big integers in memory as base $10,000$ numbers with $10,000$ digits. We will simply use standard binary to represent the digit with each word in memory being one digit of the number. When we want to output a big integer, then, we only need output each signed 32-bit integer as a group of four decimal digits, again, because our base is $10^4 = 10,000$. Similarly, when getting a number from a user, we will take the input, as a string of ASCII characters, and group it in sets of four characters to form the digits of our base $10,000$ number in memory.

The previous paragraph contains a subtle piece of information. It implies that our big integers will be stored in memory in big-endian format. This means that the if we use an array of signed integers to hold the digits that the first integer in that array will be the largest power of the base, not the smallest as it would be in little-endian format. This choice again simplifies our input and output operations.

At this point, we are able to write down how we will represent big integers for the examples that follow. We will use arrays of signed integers to represent the digits. Additionally, the very first element of the array will contain the number of digits in the number and the sign of this element will be the sign of the integer. Pictorially,

which illustrates how the number "$-12,345,678,910$" would be stored in memory using an array of four signed 32-bit integers. The first element of the array contains "-3" which says there are three digits in this number and that it is negative. The remaining three elements of the array are the digit values themselves in base $10,000$. This simple format is what we will be using going forward in this chapter.

Input and Output. With all of these preliminaries in place, Fig. 4.1 shows a simple C routine to convert a string into a base $10,000$ number. Similarly, Fig. 4.2 shows a C routine for output of a base $10,000$ number as a decimal number.

Figure 4.1 accepts a string in s and returns a pointer to a newly allocated big integer. Note that this memory must be released by a call to `free`. Let's look in detail at what this particular function is doing.

Line 10 checks to see if the input is negative. If it is we set the sign to 1 (default is 0) and skip past the sign character by incrementing our character index j. Line 15 calculates the number of digits that will be in the big integer. The input string is in base 10 while the output integer is in base $10,000$ so we divide the length of the input, accounting for the sign character, by four to see how many digits we need. We take the ceiling of this to avoid truncating the length. Line 16 allocates new memory on the heap for the number, using n+1 to add in one element for the length and sign. For clarity we skip any check on the return value of `malloc` to see if the allocation succeeded. In line 17 the sign and length are set.

Lines 19 through 36 work through the big integer, digit by digit, using sets of four characters from s to set the value. The first digit, which is the highest order digit, is a special case and is handled by lines 21 through 29. This is because the highest order digit is the only digit which might have fewer than four digits in the element. The digit is set to zero, then the `switch` statement determines how many excess digits there are in the input string. These are digits that will be in the first output digit and there can be between one to four of them. Notice there are no `break` statements in the `switch`. This will cause the first matched condition to fall through to all the lower ones. This is what we want. Lines 31 through 34 handle all other digits of the output big integer. In this case, since we know there are actually four characters matched to this digit we do not need the `switch` statement and the logic is easier to see. We successively add in each character multiplied by the proper power of

```
 1 #include <stdio.h>
 2 #include <stdlib.h>
 3 #include <math.h>
 4 #include <string.h>
 5
 6 signed int *bigint_input(char *s) {
 7     int i, j=0, n, sign=0;
 8     signed int *b;
 9
10     if (s[0] == '-') {
11         sign = 1;
12         j++;
13     }
14
15     n = (int)ceil((strlen(s)-sign) / 4.0);
16     b = (signed int *)malloc((n+1)*sizeof(signed int));
17     b[0] = (sign) ? -n : n;
18
19     for(i=1; i <= n; i++) {
20         if ((i==1) && (((strlen(s)-sign)%4) != 0)) {
21             b[i] = 0;
22             switch ((strlen(s)-sign)%4) {
23                 case 3:
24                     b[i] += (s[j++]-'0') * 100;
25                 case 2:
26                     b[i] += (s[j++]-'0') * 10;
27                 case 1:
28                     b[i] += (s[j++]-'0');
29             }
30         } else {
31             b[i]  = (s[j++]-'0') * 1000;
32             b[i] += (s[j++]-'0') * 100;
33             b[i] += (s[j++]-'0') * 10;
34             b[i] += (s[j++]-'0');
35         }
36     }
37
38     return b;
39 }
```

Fig. 4.1 Big integer input in C. This routine takes a string representing a big integer in decimal and converts it to a base $10,000$ number. The required standard C libraries are also listed

ten and converted to a digit value by subtracting the ASCII value for zero. When the loop over output digits completes the number is complete and a pointer to the memory is returned in line 38.

Figure 4.2 takes a pointer to a big integer in b and a pointer to an output string buffer in s. The big integer is then put into s as an ASCII string. This conversion is simpler than going the other way as in Fig. 4.1. If the first element of the big integer is negative the number is negative so we output the sign in line 8 bumping the output character index, j. Lines 10 through 16 pass over the big integer, digit by

```
1 #include <stdio.h>
2
3 void higint_output(signed int *b, char *s) {
4     int i, j=0;
5     char t[5];
6
7     if (b[0] < 0)
8         s[j++] = '-';
9
10    for(i=1; i <= abs(b[0]); i++) {
11        sprintf(t, "%04d", b[i]);
12        if (!((t[0] == '0') && (i == 1))) s[j++] = t[0];
13        if (!((t[1] == '0') && (i == 1))) s[j++] = t[1];
14        if (!((t[2] == '0') && (i == 1))) s[j++] = t[2];
15        s[j++] = t[3];
16    }
17
18    s[j] = '\0';
19 }
```

Fig. 4.2 Big integer output in C. This routine takes a base $10,000$ big integer and outputs it as a decimal string

digit, and convert each digit to a set of up to four ASCII characters. To save space we use a temporary string, t, and the C library function sprintf which prints into a string. This changes the i-th big integer digit into a four character string. This string, because of the %04d format statement may have leading zeros. If i is one, not zero because we skip the sign/length element, then we do not want leading zeros. This is the origin of the somewhat cryptic code in lines 12 through 14. Each of these if statements asks if we are working with the highest order digit, when i is one, and if that digit is zero. It then negates this to execute when we are not working with the highest order digit and the digit is zero. In that case, we want to copy the proper character from the temporary string, t, to the output string, s. Since we always have at least one character in the digit we are working with line 15 has no if statement, it is always executed. When the loop in lines 10 through 16 completes we have filled in all of the output string, character by character, with the decimal representation of the input big integer. All that remains is to add a null at the end of the output to make a valid C-style string out of it (line 18). With this the conversion is complete.

Comparisons. Before we jump into arithmetic routines for big integers we need to take a quick look at comparisons using big integers.

Figure 4.3 shows basic big integer comparison in C. The function bigint_compare takes two big integer arguments, a and b, and returns -1 if $a < b$, 0 if $a = b$, and $+1$ if $a > b$. Let's take a look at what it is doing.

Lines 4 and 5 do a quick comparison to see if the signs of the numbers are different. If they are we already know the proper value to return and we do not need to look at the actual data values. Lines 7 through 10 do similar quick tests. We already know that the signs of a and b are the same so we look to see which has a larger number of digits. In line 7 we look to see if a has more digits than b and that they

```
1 int bigint_compare(signed int *a, signed int *b) {
2     signed int i;
3
4     if ((a[0] > 0) && (b[0] < 0)) return 1;
5     if ((a[0] < 0) && (b[0] > 0)) return -1;
6
7     if ((abs(a[0]) > abs(b[0])) && (a[0] > 0)) return 1;
8     if ((abs(b[0]) > abs(a[0])) && (a[0] > 0)) return -1;
9     if ((abs(a[0]) > abs(b[0])) && (a[0] < 0)) return -1;
10    if ((abs(b[0]) > abs(a[0])) && (a[0] < 0)) return 1;
11
12    if (a[0] > 0) {
13        for(i=1; i <= abs(a[0]); i++) {
14            if (a[i] > b[i]) return 1;
15            if (b[i] > a[i]) return -1;
16        }
17    } else {
18        for(i=1; i <= abs(a[0]); i++) {
19            if (a[i] > b[i]) return -1;
20            if (b[i] > a[i]) return 1;
21        }
22    }
23
24    return 0;
25 }
```

Fig. 4.3 Big integer comparison. The function compares two big integers, a and b, and returns -1 if $a < b$, 0 if $a = b$, and $+1$ if $a > b$. With this basic comparison function it is straightforward to define $<, >, =, <=, >=$ and $! =$ via small wrapper functions

are both positive. If so, we know that a is larger so we return 1. If both are positive and b has more digits, the case in line 8, we return -1 because that means that a is the smaller value and we are determining the relationship between a and b (not b and a). Lines 9 and 10 do the same comparison but for both a and b negative. In this case, the return values are flipped since if a has more digits than b it must be larger in magnitude and negative making it smaller than b.

Line 12 looks at the sign of a. At this point we know a and b have the same sign and same number of digits. We must therefore look digit by digit, largest power of 10,000 first, to see where they might be different. This tells us which of the two is larger. The loop in lines 13 through 16 moves through a and b and uses digit values to decide if one is larger than the other. The loop in lines 19 through 21 does the same but in this case we know that a and b are both negative and have the same number of digits so the return values are reversed. Lastly, if we reach line 24 we know that the numbers have the same sign, same number of digits, and same digit values so we know that they must be equal and we return 0.

The function in Fig. 4.3 makes it easy to define wrapper functions to implement the actual comparison operations. If we were using C++ we would define a big integer class and overload $<, >, =$, etc to implement the comparisons. As we are using plain C we instead implement a function library. A possible definition for the equal function is,

```
int bigint_equal(signed int *a, signed int *b) {
    return bigint_compare(a,b) == 0;
}
```

We leave the definition of the remaining operators as an exercise.

Now that we know how to store big integers, how to read them from strings, how to write them to strings, how to compare them, we are ready to look at basic arithmetic with big integers.

4.3 Arithmetic with Big Integers

Basic arithmetic with big integers generally follows the school book methods though as we will see there are some alternatives for multiplication that are effective if the number of digits is very large. We will continue with our implementation of big integers as above and develop routines for addition, subtraction, multiplication and division. We will not implement C versions of the alternative multiplication routines but we will discuss the operation of the algorithms. Note that the routines given here are bare-bones only. We are omitting necessary checks for memory allocation, etc., in order to make the routines as concise and easy to follow as possible. We proceed then with the understanding that a true big integer package would have all of these checks in place.

Addition and Subtraction. Before we look at implementing addition and subtraction let's think a bit about how we will handle negative numbers. We are using a sign-magnitude representation of our big integers because using a complement notation might mean manipulating all the digits of the number when we can be more clever about handling operations with regard to the signs. For addition we have the following situations in Table 4.1 where we always work with the absolute value of the integers a and b. Similarly for subtraction we have Table 4.2 where it becomes apparent that for addition and subtraction we need the ability to compare the absolute value of two big integers, the ability to negate a big integer, and the ability to add and subtract two positive big integers. Therefore, to properly implement addition and subtraction we need two helper functions: bigint_ucompare to compare the absolute value of two big integers, and bigint_negate to negate a big integer. These are given in Fig. 4.4.

These functions are straightforward. Indeed, bigint_negate is about as simple as can be given the efficient way we have selected to represent our big integers in memory. To change the sign we need only negate the first element of the array. In addition we return a reference to the array so that we can compose the negate call with other big integer operations. The function bigint_ucompare is a reduced version of bigint_compare as given in Fig. 4.3 where we do not pay attention to the sign. It returns -1 if $|a| < |b|$, 0 if $|a| = |b|$, and $+1$ if $|a| > |b|$. With these in hand, we can implement addition and subtraction using low-level functions that ignore the

sign of a	sign of b	Steps to find a + b
+	+	Add $|a|$ and $|b|$.
+	−	$|a| > |b|$, subtract $|b|$ from $|a|$.
		$|a| < |b|$, subtract $|a|$ from $|b|$. Negate answer.
−	+	$|a| > |b|$, subtract $|b|$ from $|a|$. Negate answer.
		$|a| < |b|$, subtract $|a|$ from $|b|$.
−	−	Add $|a|$ and $|b|$. Negate answer.

Table 4.1 Steps to add two big integers paying attention to their signs

sign of a	sign of b	Steps to find a − b
+	+	$|a| \geq |b|$, subtract $|b|$ from $|a|$.
		$|a| < |b|$, subtract $|a|$ from $|b|$. Negate answer.
+	−	Add $|a| + |b|$.
−	+	Add $|a| + |b|$. Negate answer.
−	−	$|a| \leq |b|$, subtract $|a|$ from $|b|$.
		$|a| > |b|$, subtract $|b|$ from $|a|$. Negate answer.

Table 4.2 Steps to subtract two big integers paying attention to their signs

```
1  signed int *bigint_negate(signed int *a) {
2      a[0] = -a[0];
3      return a;
4  }
5
6  int bigint_ucompare(signed int *a, signed int *b) {
7      signed int i;
8
9      if (abs(a[0]) < abs(b[0])) return -1;
10     if (abs(a[0]) > abs(b[0])) return 1;
11
12     for(i=1; i <= abs(a[0]); i++) {
13         if (a[i] > b[i]) return 1;
14         if (b[i] > a[i]) return -1;
15     }
16
17     return 0;
18 }
```

Fig. 4.4 Big integer helper functions to compare the absolute value of two big integers, bigint_ucompare, and to negate a big integer, bigint_negate

sign of their arguments. After that, we will use a higher level version of addition and subtraction that does look at the sign of the arguments to decide how to proceed.

Our low-level addition routine will pay no attention to the sign of the arguments. It will assume they are positive. This routine implements addition using the simple method we learned in school. We add digits, from right to left, carrying on every 10,000. We will add a new first digit to the output, if necessary, based on any carry in the top digit position. This is directly analogous to addition in any other base. We call the routine bigint_uadd and it is given in Fig. 4.5.

```
 1|signed int *bigint_uadd(signed int *a, signed int *b) {
 2|    signed int *c, *x, *y, *t;
 3|    int i, m, n, yy, carry=0;
 4|
 5|    if (abs(a[0]) > abs(b[0])) {
 6|        n = abs(a[0]);   m = abs(b[0]);
 7|        x = a;           y = b;
 8|    } else {
 9|        n = abs(b[0]);   m = abs(a[0]);
10|        x = b;           y = a;
11|    }
12|
13|    c = (signed int *)malloc((n+1)*sizeof(signed int));
14|    c[0] = n;
15|
16|    for(i=n; i>0; i--) {
17|        yy = (i>(n-m)) ? y[i-(n-m)] : 0;
18|        c[i] = x[i] + yy + carry;
19|        if (c[i] > 9999) {
20|            carry = 1;
21|            c[i] -= 10000;
22|        } else
23|            carry = 0;
24|    }
25|
26|    if (carry) {
27|        t = (signed int *)malloc((n+2)*sizeof(signed int));
28|        for(i=1; i<=n; i++)   t[i+1] = c[i];
29|        t[0] = n+1;
30|        t[1] = 1;
31|        free(c);
32|        c = t;
33|    }
34|    return c;
35|}
```

Fig. 4.5 Low-level big integer addition using the school method

Recall that we will be using a high-level driver routine to decide when to call the low-level add and subtract routines. Because of this, bigint_uadd will ignore the sign of its input and treat all arguments as positive. Lines 5 through 11 set up the addition. We use extra pointers, x, and y, and assign them the arguments a or b so that x will always point to the larger number (line 5). We also use n and m to hold the number of digits in these numbers. We do this to align the digits properly when we move from right to left across the number adding as we go. Since y will always have as many or fewer digits as x we can use this fact in the addition to substitute zero for digits that do not exist in y. Line 13 allocates memory for the addition while line 14 sets the length of the output. We know that when adding two big integers it is possible that we overflow in the most significant digit. If that happens, we adjust for it later in the routine and initially assume that the answer will fit in n digits which is the number of digits in the larger of a or b. Lines 16 through 24 perform the actual addition, digit by digit, from the least significant digit to the most significant. We use yy to hold the digit value of the lesser of a and b (in terms of number of digits). Line 17 either returns the actual digit of y or zero if we are looking at digits of x that have no match in y. Line 18 sets the output digit in c. We add the current digit of x and y and any carry that came over from the previous digit addition. Note that we set carry to zero in line 3 when it was declared.

After the addition of the digits we need to check if we caused a carry. We are using base $10,000$ so we know that any two digits added together will still fit in a single signed int. We simply need to see if the result is larger than 9999. If so, we subtract the base and set the carry for the next digit. Otherwise, we ensure that the carry is clear (line 23).

The addition loop continues until all the digits of x have been examined. At the end of this we have the answer in c but we need to do one more check. We need to check for overflow of the most significant digit of c. This is line 26. If the carry is set from the last digit addition we have overflowed the n digits we allotted to c and need to add one more most significant digit. Lines 27 through 33 do this by creating a new output array in t with n+1 digits. Line 28 copies the digits from c to t leaving room for the new most significant digit. Lines 29 and 30 set the number of digits in t and set the most significant digit to 1 since that is the carry. Lines 31 and 32 release the old memory used by c and point c to t. Line 34 returns the pointer to the new big integer which holds the sum of a and b. This addition routine can, temporarily, use double the memory because of the copy when c overflows. For our present purposes this inefficiency is tolerable and will seldom happen in practice (how often? we don't really know, but we believe it will be infrequent).

In a similar fashion, we implement the low-level subtraction routine, bigint_usub, in Fig. 4.6. This routine, like bigint_uadd, ignores the sign of its inputs. It also assumes that a is always larger, in terms of magnitude, than b. This means that we will never underflow and end up with a negative answer. Our high-level driver routine will arrange things so that this assumption is always true. Lines 5 through 8 get the number of digits of our arguments in n and m and allocate room for our answer in c which will always need n+1 digits. We set c[0] to the number of digits.

Lines 10 through 19 perform the subtraction, from least significant digit to most significant using the same trick we used in bigint_uadd to get the digit of the smaller number in y, which may be zero if we are looking at digits b does not have. Line 12 does the subtraction for the current digit and puts it in c. Note that we subtract any borrow that happened in the previous digit subtraction. We set borrow to zero when it is defined in line 3. How do we know if we borrowed? If the digit answer is negative, we need to borrow a base value ($10,000$) from the next higher digit. The check for this is in line 13. If true, borrow is set to 1 (we only ever need to borrow $10,000$) and we add the borrow into the current digit value to bring it positive. If no borrow is needed, we set borrow to zero and continue to the next most significant digit. Line 18 appears to be curious. It adds the current digit value to a single variable we are calling zero. This is set to zero when defined in line 3. We use this variable to decide if the result of the subtraction is exactly zero.

```
 1 signed int *bigint_usub(signed int *a, signed int *b) {
 2     signed int *c, *t;
 3     int l, m, n, y, borrow=0, zero=0;
 4
 5     n = abs(a[0]);
 6     m = abs(b[0]);
 7     c = (signed int *)malloc((n+1)*sizeof(signed int));
 8     c[0] = n;
 9
10     for(i=n; i>0; i--) {
11         y = ((i-(n-m)) > 0) ? b[i-(n-m)] : 0;
12         c[i] = a[i] - y - borrow;
13         if (c[i] < 0) {
14             borrow = 1;
15             c[i] += 10000;
16         } else
17             borrow = 0;
18         zero += c[i];
19     }
20
21     if ((c[1] == 0) && (zero != 0)) {
22         t = (signed int *)malloc(n*sizeof(signed int));
23         for(i=2; i<=n; i++)   t[i-1] = c[i];
24         t[0] = n-1;
25         free(c);
26         c = t;
27     }
28
29     if (zero == 0) {
30         free(c);
31         c = (signed int *)malloc(2*sizeof(signed int));
32         c[0] = 1;
33         c[1] = 0;
34     }
35
36     return c;
37 }
```

Fig. 4.6 Low-level big integer subtraction using the school method

This allows us to simplify the return value, c, in lines 29 through 34, if possible. If we are subtracting two large integers and they happen to be exactly the same value the result, without this check, would use as much memory as the arguments themselves with every digit set to zero. This check lets us save that memory and return a very simple result that is only one digit long. The new result releases the memory used by the initial result and sets up a single digit big integer that is exactly zero instead. We also check to see if the subtraction has left us with a leading zero. This is purpose of lines 21 through 27. If the leading digit is zero, but the entire value is not zero, the leading digit is removed by allocating a new value in line 22 and copying the digits from c to this value ignoring the initial zero. Once the loop in lines 10 through 19 ends the answer is in c and we return it in line 36.

Our final `bigint_add` and `bigint_sub` routines look at the signs of their arguments and proceed to call the low-level versions of add and subtract as appropriate. Figure 4.7 shows `bigint_add` while Fig. 4.8 shows `bigint_sub`.

```
 1 signed int *bigint_add(signed int *a, signed int *b) {
 2
 3      if ((a[0]>0) && (b[0]>0))
 4          return bigint_uadd(a,b);
 5
 6      if ((a[0]<0) && (b[0]<0))
 7          return bigint_negate(bigint_uadd(a,b));
 8
 9      if ((a[0]>0) && (b[0]<0)) {
10          if (bigint_ucompare(a,b) == -1)
11              return bigint_negate(bigint_usub(b,a));
12          else
13              return bigint_usub(a,b);
14      }
15
16      if ((a[0]<0) && (b[0]>0)) {
17          if (bigint_ucompare(a,b) == 1)
18              return bigint_negate(bigint_usub(a,b));
19          else
20              return bigint_usub(b,a);
21      }
22 }
```

Fig. 4.7 High-level big integer addition

Our high-level addition routine examines the signs and magnitudes of the arguments to determine how to get the proper result. The checks in the code are directly analogous to those of Tables 4.1 and 4.2. For addition, if the arguments are both positive we simply add (line 3). If the arguments are both negative, we add and make the answer negative (line 6). If the signs are mixed we need to consider the relationship between the magnitudes of the numbers. If a is positive and b is negative (line 9) we see two cases: $|a| < |b|$ which is handled in line 11 and $|a| \geq |b|$ which is handled in line 13. Similarly, for a negative and b positive (line 16) we consider cases when $|a| > |b|$ in line 17 and when $|a| \leq |b|$ in line 20.

For high-level subtraction in Fig. 4.8 we follow Table 4.2. If both a and b are positive we look at $|a| < |b|$ in line 4 or otherwise in line 7. Similarly, if both are negative we look at $|a| > |b|$ in line 11 and otherwise in line 14. The simpler cases here are if a is positive and b is negative, line 17, or a is negative and b is positive, line 20. The signs of the arguments will fit one of these cases for both addition and subtraction so we know that all possibilities have been considered. Note that for subtraction we always ensure that the first argument to `bigint_usub` is the same size or larger than the second.

```
 1 signed int *bigint_sub(signed int *a, signed int *b) {
 2
 3      if ((a[0]>0) && (b[0]>0)) {
 4          if (bigint_ucompare(a,b) == -1)
 5              return bigint_negate(bigint_usub(b,a));
 6          else
 7              return bigint_usub(a,b);
 8      }
 9
10      if ((a[0]<0) && (b[0]<0)) {
11          if (bigint_ucompare(a,b) == 1)
12              return bigint_negate(bigint_usub(a,b));
13          else
14              return bigint_usub(b,a);
15      }
16
17      if ((a[0]>0) && (b[0]<0))
18          return bigint_uadd(a,b);
19
20      if ((a[0]<0) && (b[0]>0))
21          return bigint_negate(bigint_uadd(a,b));
22 }
```

Fig. 4.8 High-level big integer subtraction

Multiplication. Big integer multiplication can be implemented using the school method without much difficulty. We do that in Fig. 4.9 where we pay no attention to the signs of the arguments. Figure 4.10 is our high-level routine that manages the signs. Let's look at Fig. 4.10 first. The rules of multiplication say that when the signs of the multiplicand and multiplier are the same, either both positive or negative, that the answer is positive. This check is made in line 3. If the signs are the same, we simply call bigint_umult as that will always give a positive result. If the signs are different, we drop to line 7 and negate the result since the result should be negative.

Figure 4.9 is really three separate functions all of which are used to multiply the numbers stored in a and b. Lines 1 through 7 define a simple helper function that duplicates an existing big integer. It allocates space for the new integer and copies all the bytes of a to b by calling the C library function, memcpy. It then returns the reference to this new big integer.

Lines 9 through 29 define bigint_umultd which multiplies a big integer in a by a single digit, d. When we multiply two numbers using the school method we accumulate partial products, shifting them for the power of the base we are currently multiplying by. We then add these together to get the final result. For example,

$$
\begin{array}{r}
1776 \leftarrow \textit{multiplicand} \\
\times \quad 1492 \leftarrow \textit{multiplier} \\
\hline
3552 \leftarrow \textit{1st partial product} \\
159840 \\
710400 \\
+ \ 1776000 \leftarrow \textit{4th partial product} \\
\hline
2649792
\end{array}
$$

```
 1  signed int *bigint_copy(signed int *a) {
 2      signed int *b;
 3
 4      b = (signed int *)malloc((abs(a[0])+1)*sizeof(signed int));
 5      memcpy((void *)b, (void *)a, (abs(a[0])+1)*sizeof(signed int));
 6      return b;
 7  }
 8
 9  signed int *bigint_umultd(signed int *a, signed int d, unsigned int s) {
10      signed int *p, carry=0, n;
11      int i;
12
13      n = abs(a[0]) + s + 2;
14      p = (signed int *)malloc(n*sizeof(signed int));
15      memset((void *)p, 0, n*sizeof(signed int));
16      p[0] = n-1;
17
18      for(i=abs(a[0]); i>=1; i--) {
19          p[i+1] = a[i] * d + carry;
20          if (p[i+1] > 9999) {
21              carry = p[i+1] / 10000;
22              p[i+1] = p[i+1] % 10000;
23          } else
24              carry = 0;
25      }
26
27      if (carry) p[1] = carry;
28      return p;
29  }
30
31  signed int *bigint_umult(signed int *a, signed int *b) {
32      int i, n;
33      signed int *m, *x, *y;
34      signed int *p=(signed int *)NULL, *q=(signed int *)NULL;
35
36      if (abs(a[0]) > abs(b[0])) {
37          x = a;  y = b;
38      } else {
39          x = b;  y = a;
40      }
41
42      m = (signed int *)malloc(2*sizeof(signed int));
43      m[0] = 1;  m[1] = 0;
44
45      for(i=abs(y[0]); i>=1; i--) {
46          if (p != NULL) free(p);
47          if (q != NULL) free(q);
48          p = bigint_umultd(x, y[i], abs(y[0])-i);
49          q = bigint_uadd(m, p);
50          if (m != NULL) free(m);
51          m = bigint_copy(q);
52      }
53
54      if (p != NULL) free(p);
55      if (q != NULL) free(q);
56      return m;
57  }
```

Fig. 4.9 Low-level big integer multiplication using the school method

```
1  signed int *bigint_mult(signed int *a, signed int *b) {
2
3      if (((a[0]>0) && (b[0]>0)) ||
4          ((a[0]<0) && (b[0]<0)))
5          return bigint_umult(a,b);
6
7      return bigint_negate(bigint_umult(a,b));
8  }
```

Fig. 4.10 High-level big integer multiplication

We can get the same result by starting with a product of zero and adding each digit times the multiplicand after shifting to the left as many places as the location of the digit in the multiplier. This is exactly what we are doing when we write and then sum the partial products. This same approach works regardless of base. In bigint_umultd we do the same thing. The multiplicand is in a, the digit to multiply by is in d and the number of places to shift the result to the left is in s. In Line 13 of Fig. 4.9 n is set to the number of output digits. We need to have room for all the digits in a, which we get with abs(a[0]). To this we add s, these are extra digits at the end of the number that we will initialize to zero. This takes the place of the zeros added to the end of the partial products in the example above. We need to add one more array element to hold the sign and length of the number and to this we add still one more digit to handle any overflow (carry in the last digit multiplied). This is the origin of the + 2 in line 13. Line 14 allocates the product and points p to it. Line 15 erases the entire number and sets it to zero while line 16 sets the length of the number. This initializes the partial product.

Lines 18 through 25 do the multiplication by the digit, d. We look at all the digits of the multiplicand a from least significant to most significant. This defines the loop in line 18. Line 19 does the multiplication of the digit of a by d. There are two things to note in this line. First, we add in carry, just as we did for addition, but in this case the carry may be more than one. Second, we assign this result to p[i+1] and not p[i]. This is done to leave a leading zero in p to handle any final overflow. Recall that array elements further from p[0] are less significant. Lines 20 and 24 handle any carry. If the resulting multiplication is greater than 9999 we have a carry. Unlike addition, the carry is not simply one but is the value of p[i+1] divided by the base. The actual value that should be in p[i+1] is the remainder when dividing by the base. If the carry has not happened, line 24 clears carry.

When the loop exits it is possible that carry is not zero. Line 27 checks for this and sets the leading zero already allocated to the carry. Line 28 returns the partial product. A full implementation would do something about the leading zero in p, but we ignore it here.

The function that does the actual unsigned multiplication is bigint_umult. The result of the multiplication will be in m which we initialize to zero in lines 42 and 43 of Fig. 4.9. Before this, lines 36 through 40 set pointers x and y so that x always points to the input with the largest number of digits, which may be the same as the

other input. We do this because we need fewer passes through the loop in line 45 if
we make the multiplier the value with the least number of digits.

The loop in lines 45 through 52 accumulates the partial products of the digits of
y (the shorter of a and b) with x. This code is inefficient because it requires copying
and free-ing memory more than is necessary but this makes the algorithm easier to
follow. We use p and q to point to temporary results. In this case, p points to the
partial product returned by the call to bigint_umultd. This call multiplies x by
the current digit of y and shifts the result abs(y[0])-i places to the left. As we
move from right to left over the digits of y we need to shift each partial product the
current digit position to the left and this is exactly abs(y[0])-i. When this call
finishes the partial product is in p and it must be added to our accumulated product.
We reuse bigint_uadd to do this and store the new running product value in q. We
really want this result in m so we release memory used by m and call bigint_copy
to copy q to m. This is necessary so that the next pass through the loop can update q.
This is a consequence of using plain C and not a language with automatic garbage
collection. At the end of the loop we have the final product in m. To avoid leaking
memory we must free p and q at the end of the loop. Finally, we return the product
in line 56.

Clearly this implementation is pedagogical as there are more efficient ways to
implement the multiplication routine. The school approach is well known to be of
order n^2 which is denoted $O(n^2)$. Below we will discuss alternate big integer multiplication routines that do better than this. If a routines runs in $O(n^2)$ time it means
that as n, here the number of the digits, increases the running time increases as the
square of the number of digits (with an unstated multiplier that is generally ignored).
An $O(n^2)$ routine may be perfectly acceptable for small inputs (relatively few digits)
but quickly becomes unusable when the number of digits becomes large.

Division. Division of big integers is rather difficult to implement quickly. The
school algorithm for division is typically implemented but for the purposes of this
book even that is too long. So, in order to provide a minimalist but complete package
for big integer arithmetic we will instead implement the simplest approach possible.
Fear not, however, we will discuss more comprehensive algorithms for division,
along with multiplication, in the next section.

So, how to implement division with big integers? Thinking back to the simplest
possible ways of working with numbers we remember that division of two integers is
really counting the number of times the divisor can be subtracted from the dividend
before we hit zero or go below zero if the division would leave a remainder. In this
case, we ignore the remainder and leave implementing it as an exercise

As with multiplication, we implement division as a high-level function that handles the signs of the numbers and a low-level function that deals with the magnitude
only. Figure 4.11 shows our "division by repeated subtraction" approach. We divide
a by b, ignoring signs, by counting the number of times we can subtract b from a,
updating a in the process, before we hit zero or go negative. This is the quotient
which we store in q. Lines 4 through 6 initialize the quotient and set it to zero. Likewise, lines 8 through 10 initialize a constant value of 1 which is added to q each
time through the loop.

```
 1│signed int *bigint_udiv(signed int *a, signed int *b) {
 2│    signed int *q, *t, *c, *x=NULL, *y=NULL;
 3│
 4│    q = (signed int *)malloc(2*sizeof(signed int));
 5│    q[0] = 1;
 6│    q[1] = 0;
 7│
 8│    c = (signed int *)malloc(2*sizeof(signed int));
 9│    c[0] = 1;
10│    c[1] = 1;
11│
12│    t = bigint_copy(a);
13│
14│    while (1) {
15│        if (x != NULL) free(x);
16│        if (y != NULL) free(y);
17│        x = bigint_usub(t,b);
18│        y = bigint_uadd(q,c);
19│        free(t);
20│        free(q);
21│        t = bigint_copy(x);
22│        q = bigint_copy(y);
23│
24│        if (bigint_ucompare(t,b) < 1)
25│            break;
26│    }
27│
28│    free(x);
29│    free(y);
30│    free(t);
31│    free(c);
32│    return q;
33│}
```

Fig. 4.11 Low-level big integer division

Line 12 copies a so we can subtract from it without altering it in memory. The loop of lines 14 through 26 do the work. We need to manage memory carefully to avoid leaks so we use auxiliary pointers x and y to hold intermediate results. Line 17 subtracts the divisor in b from t which is the portion of a remaining. Line 18 bumps the quotient counter by adding the constant 1 in c. Lines 21 and 22 copy the partial results of the loop to t and q. We do this so we can free memory before we lose the pointer address. Languages with automatic garbage collection can avoid this sort of inelegance. Line 24 checks to see if we have reached a point where the remainder in t is less than b, the divisor. If so, we are done and break out of the loop. After cleaning up memory the quotient is returned.

It is well-known that this approach to division, while correct, is terribly ineffi-cient, especially for fixed length integers. Be that as it may, the algorithm is straight-forward and simple to understand. All we need do now is account for the signs of the dividend and divisor. This is done in Fig. 4.12 through the `bigint_div` routine.

```
 1 signed int *bigint_divide(signed int *a, signed int *b) {
 2     int sa, sb;
 3     signed int *c;
 4
 5     if ((bigint_iszero(b)) ||
 6         (bigint_ucompare(a,b) == -1)) {
 7         c = (signed int *)malloc(2*sizeof(signed int));
 8         c[0] = 1;
 9         c[1] = 0;
10         return c;
11     }
12
13     sa = (a[0] < 0) ? -1 : 1;
14     sb = (b[0] < 0) ? -1 : 1;
15     a[0] = abs(a[0]);
16     b[0] = abs(b[0]);
17
18     if (((sa<0) && (sb<0)) ||
19         ((sa>0) && (sb>0))) {
20         c = bigint_udiv(a,b);
21     } else {
22         c = bigint_negate(bigint_udiv(a,b));
23     }
24
25     a[0] = sa*a[0];
26     b[0] = sa*b[0];
27
28     return c;
29 }
```

Fig. 4.12 High-level big integer division

In order to handle the signs of a and b it is necessary to check several conditions. Along the way we check for situations that result in invalid outputs for the case of integer division or division by zero. Lines 5 through 11 do this. If the divisor is zero (bigint_iszero(b) == 0) or the dividend is less than the divisor, we return a constant value of zero. For the second case this is correct but division by zero should instead signal an error. For clarity, we ignore the error and return the incorrect value of zero instead. This would be corrected if implementing a full big integer package.

We extract the signs of the arguments in lines 13 and 14. We then ensure that they are positive so we can pass the values to bigint_udiv. We will reset the signs of the arguments when done, see lines 25 and 26. We know from school that if the signs of a and b are the same, either both positive or both negative, that the quotient must be positive. So, if this is the case, we execute line 20 to perform the division. If the signs of the arguments are opposite, either a positive and b negative or vice versa, we must make the answer negative as well. In this case, line 22 is what gets executed. After restoring the proper signs of the arguments we return the quotient in line 28.

Division by repeated subtraction is unsatisfying in the long run and can be incredibly inefficient. However, it completes our implementation of basic arithmetic for big integers and therefore is worth looking at even if it is not how things would be done if we were developing a big integer package for wide-spread use.

The algorithms we implemented above are adequate for simple purposes but, especially for multiplication and division, people have certainly done far better. In the next section we look at some of these improved algorithms without implementing them. This will give an appreciation for what goes into efficient algorithm implementation and design and will give us a background we will use when discussing particular big integer libraries that you may someday wish to use in your own programs.

4.4 Alternative Multiplication and Division Routines

In this section we take a look at some alternate algorithms for multiplication and division. A true big integer package would implement some if not all of these alternatives. By necessity we are cherry-picking algorithms. A particularly nice treatment of big integer algorithms is found in Chapter 9 of Crandall and Pomerance [1] and readers interested in a more thorough mathematical treatment are directed there.

Multiplication. In the previous section we implemented the school method for multiplication of big integers. As we noted, this method runs $O(n^2)$ in the size of the inputs. While this is adequate for numbers with perhaps a thousand digits, it becomes intractable if we are working with very large numbers with tens of thousands, hundreds of thousands or even millions of digits. To work with truly large big integers we need to improve our multiplication algorithm. Fortunately, this problem has been investigated and a number of alternate multiplication algorithms exist. Here we will look at three of them: Comba multiplication, Karatsuba multiplication and FFT multiplication. Each has an advantage over the simple school method and with respect to each other.

Our first possible improvement to multiplication is the Comba algorithm [2]. Comba noted that the school method of multiplication is geared towards humans and not computers. After each single digit multiplication of a column we carry to the next column. In Comba multiplication we instead do not carry but store the entire product for each column position. We then repeat for the next digit to multiply adding in any existing value in that column (from a previous digit multiplication). When all single digit multiplications have been performed the resulting vector of column values is adjusted from left to right by adding any carry to the next digit to the left and leaving the remainder in the column. When this is done the vector will contain the output value. This process is described in Fig. 4.13 where we work in base 10 for clarity. If we were implementing this algorithm for our big integer representation we would be using base $10,000$ instead.

			2	3	7	
×			1	4	8	
			16	24	56	← *237×8*
a		8	12	28		← *237×4*
	+ 2	3	7			← *237×1*
	2	11	35	52	56	← *column sums*
	3	5	0	7	6	← *carries, right to left*

			2	3	7	
×			1	4	8	
b			16	24	56	← *column1*
		8	28	52	56	← *column2*
	2	11	35	52	56	← *column3*
	3	5	0	7	6	← *carries, right to left*

Fig. 4.13 Comba column-wise multiplication of 237×148 in base 10. *(a)* Writing out the multiplication by hand keeping each column without carries to the next. After all digits have been multiplied we add the columns. Finally, we perform the carries from right to left. *(b)* The same but as would be implemented in a computer. In this case, a single output number stores the columns adding the new value for that column to the previous. Once all the digits of the multiplier have been visited the output is adjusted by carrying from right to left to achieve the same answer as in *(a)*

In *(a)* we work out the multiplication using the school method but instead of carrying after each multiplier digit multiplication we keep the value in the proper column. When each digit of the multiplier has been used we sum the results to get an intermediate answer that still needs to be adjusted for the carries we did not perform in the previous steps. The last step is to move from right to left replacing each column value, n, with $n\%10$, and adding to the next digit the result of $n/10$ using integer division. This handles all the carries we did not do in the lines above. In *(b)* we do the same as in *(a)* but each time we multiply a digit we immediately add the result to the proper column of the single output value. Then, when done, we handle the carries.

Comba multiplication is generally believed to be about 30% more efficient than the school method. However, there are some things to keep in mind before using it. Depending upon the way digits of the big integer are stored it is possible to overflow a particular digit value because a column may need to sum the single digit multiplication for n digits where n is the minimum number of digits in the multiplicand and multiplier. For that reason, Comba multiplication is limited in the number of digits in the product and is not suitable for multiplying truly large numbers. Additionally, while a 30% improvement is helpful, the algorithm is still essentially the school method and therefore is still $O(n^2)$. We can do better than this.

Karatsuba multiplication [3] improves on the school method by simplifying the multiplication. The essential insight involves simplifying the multiplication of two large integers, a and b, by replacing the multiplication with three multiplications of numbers with about half as many digits as a and b. If we choose m, $m < n$, where n is the number of digits, we can rewrite a and b as,

$$a = a_1 B^m + a_0$$
$$b = b_1 B^m + b_0$$

where a_0, a_1, b_0, and b_1 are all less than B^m where B is the base of our numbers, for example, $B = 10,000$. If we then calculate ab we get,

$$ab = (a_1 B^m + a_0)(b_1 B^m + b_0) = a_1 b_1 B^{2m} + (a_1 b_0 + a_0 b_1)B^m + a_0 b_0$$

which we can rewrite as,

$$\alpha B^{2m} + \beta B^m + \gamma$$

with,

$$\alpha = a_1 b_1$$
$$\beta = a_1 b_0 + a_0 b_1$$
$$\gamma = a_0 b_0$$

This is seemingly no improvement. Instead of the single multiplication ab we now have *four* multiplications to get α, β and γ, but press on anyway and we will soon see the benefit. We can remove one of the four multiplications by rewriting our expressions. First we look at $(a_1 + a_0)(b_1 + b_0)$,

$$(a_1 + a_0)(b_1 + b_0) = a_1 b_1 + a_1 b_0 + a_0 b_1 + a_0 b_0$$
$$= (a_1 b_0 + a_0 b_1) + a_1 b_1 + a_0 b_0$$

where the first term on the right is β. This means we can determine β with one less multiplication,

$$a_1 b_0 + a_0 b_1 = (a_1 + a_0)(b_1 + b_0) - a_1 b_1 - a_0 b_0$$
$$\beta = (a_1 + a_0)(b_1 + b_0) - \alpha - \gamma$$

One thing to note is that $(a_1 + a_0)(b_1 + b_0)$ can itself be calculated using Karatsuba multiplication meaning this algorithm is recursive. This is one of its strong suits.

So, where is the benefit? It comes from the fact that the three multiplications we have replaced ab with are numbers with roughly half the number of digits as a or b. While a formal complexity analysis is beyond the scope of this book we see that the additions and subtractions are linear, $O(n)$, and as n grows they become less important. The recursive nature makes the analysis a bit tricky but the final result is that the algorithm as a whole is $O(n^{lg(3)})$ where $lg(3)$ is the log base 2 of 3. Compare this to the $O(n^2)$ of the school method and as Fig. 4.14 shows, this difference quickly becomes important for large n.

If n is too small, say less than four digits, it is more efficient to use the school method to multiply ab. This means we can use the school method as the terminating condition of the recursive application of Karatsuba multiplication.

Our final multiplication algorithm is that of Schönhage and Strassen [4]. This algorithm is based deeply in abstract algebra and Fourier transform theory but we

Fig. 4.14 $O(n^{lg(3)})$ (*solid line*) versus $O(n^2)$ (*dashed line*) growth as n increases. This plot shows that the Karatsuba method is a significant improvement over the school method for suitable input lengths (n)

will describe it at a high level. Here we discuss the variant presented as Algorithm 9.5.23 in [1].

Before diving into the algorithm itself we need to address a few concepts related to the operation of the algorithm which are its connection to abstract algebra and a bit about convolution, multiplication, and Fourier transforms.

The algorithm actually calculates multiplication modulo a specific integer, $2^n + 1$. This means that the result, which is an integer, is really a number from $\mathbb{Z}_{2^n+1}$. If you are not familiar with the notation, $\mathbb{Z}$ refers to the set of integers and $\mathbb{Z}_m$ refers to the set of integers modulo m. For example, $\mathbb{Z}_4 = 0, 1, 2, 3$ and operations that go above or below "wrap around" as needed so that in $\mathbb{Z}_4$ we have $2 + 3 = 1$. This is not a problem, really, since we select n so that $2^n + 1$ is larger than the product of our arguments. Call the arguments x and y so we select n such that $2^n + 1 > xy$. In this case, we never "wrap around" and we get the desired product.

In reality, the algorithm operates on a *ring* [5] defined in $\mathbb{Z}_{2^n+1}$ and uses only integer operations. A ring is an enhancement of a group which is itself a generalization of a binary operation acting on a set. In a ring, along with the first binary operation of the group, which is often denoted as $+$ even though it need not be addition, we add a second operation, $\bullet$, which need not really be multiplication. This is the abstract algebra connection. In our case, addition is addition, multiplication is multiplication and the set is $\mathbb{Z}_{2^n+1}$, which is a (possibly very large) set of integers. If we select n so that $n \geq \lceil \lg x \rceil + \lceil \lg y \rceil$ then $2^n + 1 > xy$ which is what we want so that the algorithm produces the actual product.

In the Comba algorithm for multiplication we saw that multiplying the multiplicand by a digit of the multiplier, especially when we collected the resulting value without carrying, was in essence doing convolution. A convolution is passing one value over another, so here we "pass" the single multiplier digit "over" the multiplicand digits, and collect the product. This means that multiplication of two numbers is related to convolution. This is where the Fourier connection comes from. We need not know anything about Fourier transforms at present beyond the fact that they map a sequence of values from one space to another space. We denote a Fourier transform of x, where x is a vector of values, by $\mathscr{F}(x)$ and recognize that this forward transform has an inverse transform, $\mathscr{F}^{-1}$, so that $\mathscr{F}^{-1}(\mathscr{F}(x)) = x$. In truth, the Fourier transform is such a powerful mathematical tool that all readers should be familiar with it. Please take a look at Bracewell [6] or a similar text to familiarize yourself with this elegant addition to the mathematical toolbox.

One of the many properties of the Fourier transform is that multiplication in one space is equivalent to convolution in another. This is a key hint in understanding what the Schönhage and Strassen algorithm is doing. It will calculate the convolution of x and y through the Fourier transform and multiplication when in Fourier space. In essence, the algorithm does,

$$X' = \mathscr{F}(X), Y' = \mathscr{F}(Y) \leftarrow \text{map to Fourier space}$$
$$Z' = X'Y' \qquad\qquad\quad \leftarrow \text{calculate a product}$$
$$Z = \mathscr{F}^{-1}(Z') \qquad\quad\; \leftarrow \text{return to } X \text{ and } Y \text{ space}$$

which at first seems backwards but isn't because the result in z is the convolution of X and Y which is what we want. It is important to note that X and Y are not x and y, the numbers to be multiplied, but decompositions of them as performed in the first part of the algorithm. We are intentionally glossing over the fact that the Fourier transform referred to by $\mathscr{F}$ is not the one familiar to scientists and engineers but a generalization of it that operates via integer operations only. See Definition 9.5.3 of [1], particularly the integer-ring DFT variant. We are now ready to look at the Schönhage and Strassen algorithm which we present in Fig. 4.15 with apologies to Crandall and Pomerance for the paraphrasing.

In step 1 we select a size for our Fourier transforms and n'. The selection of n' is not arbitrary. It is selected so that the algorithm will work on smaller sets in step 5 should the multiplication there, which is also of the form $xy \bmod N$ where $N = 2^{n'} + 1$, be done by a recursive call to the algorithm itself. In step 2 we then split the input values, x and y, into vectors A and B. It is these vector representations that will be mapped to Fourier space to perform the convolution. Recall that the convolution in this space will mimic the school method of multiplication.

In Step 3 we weight the vectors A and B to prepare for the FFT, remembering that the FFT here implements integer-only operations. The multiplication can be performed using only shifts in a binary representation of the numbers. Also, the modulo operation can likewise be implemented efficiently as Crandall and Pomerance also show in [1].

Step 4 performs the actual FFT in-place to save memory as these numbers may be very large. Step 5 multiplies the components in Fourier space, again this is convolu-

1. [Initialize]
   Choose an FFT size, $D = 2^k$ dividing n.
   Set a recursion length, $n' \geq 2M + k$, $n = DM$ such that D divides n', $n' = DM'$.
2. [Decomposition]
   Split x and y into arrays A and B of length D such that $A = A_0, A_1, \ldots, A_{D-1}$ and $B = B_0, B_1, \ldots, B_{D-1}$ as residues modulo $2^{n'} + 1$.
3. [Prepare FFT]
   Weight A and B by $(2^{jM'} A_j) \bmod (2^{n'} + 1)$ and $(2^{jM'} B_j) \bmod (2^{n'} + 1)$ respectively for $j = 0, 1, \ldots, D - 1$.
4. [Perform FFT]
   $A = \mathscr{F}(A)$ and $B = \mathscr{F}(B)$ done in-place.
5. [Dyadic stage]
   $A_j = A_j B_j \bmod (2^{n'} + 1)$ for $j = 0, 1, \ldots, D - 1$. Storing product in Fourier space in A.
6. [Inverse FFT]
   $A = \mathscr{F}^{-1}(A)$. A now contains the convolution of the decomposition of x and y. This is the product.
7. [Normalization and Sign Adjustment]
   Scale A assigning to C to account for scale factor in Fourier transform.
8. [Composition]
   Perform carry operations to C to give final sum: $xy \bmod (2^n + 1) = \Sigma_{j=0}^{D-1} C_j 2^{jM} \bmod (2^n + 1)$ which is the desired product.

Fig. 4.15 The Schönhage and Strassen big integer multiplication algorithm as presented in [1] with paraphrasing and explanatory text added

tion in the original space, and it is this step that requires something other than shifts and basic integer operations. Notice that the form of this multiplication matches what this very algorithm does so it is possible to make a recursive call here which is the motivation for selecting n' properly. Other multiplication methods could be used here as well including the basic school method. Notice that the result is stored in A, again to save memory. Step 6 performs the inverse FFT to return to the original space and Step 7 normalizes the result by applying a scale factor that is typical of the FFT, either split between the two transforms or applied in one direction. During this process the vector C is created. Lastly, step 8 accounts for the carries (ala Comba above) and returns the result which is the desired product. A complexity analysis of this algorithm shows that it runs in $O(n \log n \log \log n)$ which is significantly better than the $O(n^2)$ of the school method.

Division. For division we look at two algorithms. The first is that found in Knuth [7] known as Algorithm D which is an implementation of long division suitable for application to big integers. The second is the divide-and-conquer division algorithm of Burnikel and Ziegler [8].

In [7] Knuth gives a thorough description his long division algorithm which follows the school method except for one twist. The twist has to do with trial divisions based on an observation, which he proves, that assuming the quotient of an $n + 1$ digit number x divided by an n digit number y to be the quotient of the top two digits of x by the leading digit of y means that your assumption will never be more than 2 away from the true value. A few examples of this fact,

$$
\begin{array}{rl|rl|l}
24356 / 5439 = & 4 & 24 / 5 = & 4 & 0 \\
95433 / 6796 = & 14 & 95 / 6 = & 15 & 1 \\
12345678 / 9876543 = & 1 & 12/9 = & 1 & 0 \\
8888888888 / 777777777 = & 11 & 88/7 = & 12 & 1
\end{array}
$$

where the right-most column is the difference between the exact quotient and the trial quotient. It is important to note, however, that this fact is only true if the divisor is greater than or equal to the one half the base, here base is 10 so the divisors are all 5 or larger. The normalization step of the algorithm takes case of this case by multiplying both dividend and divisor by a value, which can be done with a simple shift, since the quotient will be the same and the remainder can be unnormalized at the end.

The essence of the Knuth algorithm, paraphrased from [7], is given below,

1. [Normalize]
   Scale inputs u and v by a single digit, d so that the leading digit of v, called v_1, is $v_1 \geq \lfloor b/2 \rfloor$ where b is the base. Note that u and v are big-endian vectors representing the digits of the numbers in base b.

2. [Initialize]
   Loop m times where m is the difference between the number of digits in u and v. Naturally, if m is negative the quotient is 0 and the remainder is v since $v > u$. Each pass through the loop is division of the n digits of u starting at j, the loop index, by v.

3. [Calculate Trial Quotient]
   Set $\hat{q} = \lfloor (u_j b + u_{j+1})/v_1 \rfloor$, the trial quotient. If $u_j = v_1$ set $\hat{q} = b - 1$. Knuth adds a test here that will ensure that the trial quotient is never more than one greater than it should be.

4. [Multiply and Subtract]
   Subtract $\hat{q}$ times v from the n digits of u, starting at position j, replacing those digits. This is the subtraction normally done in long division by hand. Since it is possible $\hat{q}$ is too large this result may be negative. If so, it will be handled in step 6.

5. [Test Remainder]
   Set the j-th digit of the quotient to the trial value in $\hat{q}$. If the result of step 4 was negative, do to Step 6, otherwise, go to Step 7.

6. [Add Back]
   Decrease the j-th digit of the quotient by one and add v back to u starting with the j-th digit. This accounts for the few times the trial quotient is too large.

7. [Loop]
   Go back to Step 2 while $j \leq m$.

8. [Unnormalize]
   The correct quotient is in q and the remainder is u_{m+1} through u_{m+n} divided by d, the scale factor used in the first step.

The run time of this algorithm is $O(mn)$ which is adequate for smaller input sizes but becomes cumbersome if the inputs are very large in much the same way the school multiplication algorithm does.

Burnikel and Ziegler [8] implement two algorithms which work together to perform division of big integers with specific numbers of digits. They then describe how to apply these algorithms to arbitrary division problems. The first algorithm, which they name $D_{2n/1n}$, divides a number of $2n$ digits by a number of n digits. They do this by breaking the number up into super digits. If we wish to find A/B and use this algorithm we must break A into four digits, (A_1, A_2, A_3, A_4) and break B into two digits (B_1, B_2) so that each digit has a length of $n/2$. In other words,

$$A = A_1 \beta^{3n/2} + A_2 \beta^n + A_3 \beta^{n/2} + A_4$$
$$B = B_1 \beta^{n/2} + B_2$$

With this separation of A and B into super digits we can examine the $D_{2n/1n}$ algorithm,

Algorithm $D_{2n/1n}$

1. [Given]
   $A < \beta^n B$, $\beta/2 \leq B < \beta^n$, n even, β is the base. Find $Q = \lfloor A/B \rfloor$ and remainder $R = A - QB$.
2. [High Part]
   Compute Q_1 as $Q_1 = \lfloor [A_1, A_2, A_3]/[B_2, B_1] \rfloor$ with remainder $R_1 = [R_{1,1}, R_{1,2}]$ using algorithm $D_{3n/2n}$.
3. [Low Part]
   Compute Q_2 as $Q_2 = \lfloor [R_{1,1}, R_{1,2}, A_4]/[B_2, B_1] \rfloor$ with remainder R using algorithm $D_{3n/2n}$.
4. [Return]
   Return $Q = [Q_1, Q_2]$ as the quotient and R as the remainder.

This algorithm is deceptively simple. The recursive nature of it becomes clear when we look at algorithm $D_{3n/2n}$ which in turn calls $D_{2n/1n}$ again. First we present $D_{3n/2n}$ and then we discuss them together,

Algorithm $D_{3n/2n}$

1. [Givens]
   $A = [A_1, A_2, A_3]$ and $B = [B_1, B_2]$ from $D_{2n/1n}$. Find $Q = \lfloor A/B \rfloor$ and remainder $R = A - QB$.
2. [Case 1]
   If $A_1 < B_1$ compute $\hat{Q} = \lfloor [A_1, A_2]/B_1 \rfloor$ with remainder R_1 using $D_{2n/1n}$.
3. [Case 2]
   If $A_1 \geq B_1$ set $\hat{Q} = \beta^n - 1$ and set $R_1 = [A_1, A_2] - [B_1, 0] + [0, B_1] = [A_1, A_2] - \hat{Q}B_1$.
4. [Multiply]
   Compute $D = \hat{Q}B_2$ using Karatsuba multiplication.
5. [Remainder]
   Compute $\hat{R} = R_1 \beta^n + A_4 - D$.

6. [Correct]

   As long as $\hat{R} < 0$ repeat:

   a. $\hat{R} = \hat{R} + B$
   b. $\hat{Q} = \hat{Q} - 1$.

7. [Return]

   Return $\hat{Q}$ and $\hat{R}$.

The pair of algorithms, $D_{2n/1n}$ and $D_{3n/2n}$, work together on successively smaller inputs to return the final solution to the division problem. We know that $D_{2n/1n}$ requires A to have twice as many digits as B and when it is called in the first [Case 1] of $D_{3n/2n}$ we are also passing it arguments that meet the $2n$ to n digit requirement so the recursion will end.

It is interesting to compare the steps in $D_{3n/2n}$ with those of the Knuth algorithm above. We get a "trial" quotient, $\hat{Q}$, from one of the cases and then determine the remainder in step 5. Then in step 6 we correct the quotient in case we selected too large of an answer. This is exactly what is done in the Knuth algorithm in step 6. Here we are structuring the problem in such a way that the digits always match the desired pattern and using recursion to divide up the work. In the Knuth algorithm we move through the divisor digit by digit.

The school division method of Knuth is $O(mn)$ which would be $O(n^2)$ for the case we are considering here since the dividend has twice as many digits as the divisor. What about the Burnikel and Ziegler algorithm? As they discuss in [8] the performance of this algorithm depends upon the type of multiplication used in step 4 of $D_{3n/2n}$. If using Karatsuba multiplication they determine the performance to be $2K(n) + O(n \log n)$ with $K(n)$ the performance of Karatsuba multiplication itself. As they note, if the school method of multiplication is used instead we get $n^2 + O(n \log n)$ which is actually worse than long division itself. So, the choice of multiplication algorithm is crucial in determining the performance.

We have examined, superficially, several alternatives to the basic arithmetic algorithms we developed for our big integer implementation. We have seen that they out perform the naive algorithms and that they extend our abilities in working with big integers to those that are much larger than what we would be able to work with otherwise. Computer memory is inexpensive so the limiting factor is now algorithm performance. Every small improvement counts. There are many other algorithms we could have examined but cannot for space considerations. There is ongoing research in this area and tweaks and small improvements are happening all the time. But, for now, we leave the algorithms and move on to more practical considerations, namely, implementations. It is fun to implement some of these basic algorithms ourselves, but the complex and highly performant ones are often tricky and it is wise to use proven libraries when speed and accuracy count.

4.5 Implementations

Most programming languages do not define big integer operations as intrinsic to the language. In this section we will look at some that do, specifically Python and Scheme. We will also look at some extensions to existing languages such as C/C++, Java and JavaScript which give big integer functionality to developers using these languages. We start first with the extensions.

Libraries. When discussing software libraries time is naturally a factor. Projects start and finish and are often abandoned as people move on to other work making an orphan of the library which becomes less and less relevant as time moves on. Here we will look at one library that has considerable popular support and is a member of the Gnu Software suite, namely, the *GNU Multiple Precision Arithmetic Library* or GMP for short [9]. This library is open source software and is distributed via the GNU LPGL and GNU GPL licenses which allows the library to be used in its existing form in commercial code while enabling additions to the library to be passed on to other users. It has been in constant development since 1991 which is why we include it here as it is likely to be supported for some time to come.

GMP is a C and C++ library for a wide-range of mathematical operations involving multi-precision numbers. At its core it consists of several main packages: signed integer arithmetic functions (*mpz*), rational arithmetic functions (*mpq*), floating-point functions (*mpf*), and a low-level implementation (*mpn*). Additionally, a C++ class wrapper is available for the *mpz*, *mpq* and *mpf* packages. Here we will consider only the integer operations of the *mpz* package. The GMP library goes to great lengths to ensure efficiency and optimal performance.

The *mpz* package uses much the same format for numbers as we have used above. It defines a structure in gmp.h called __mpz_struct which is typedef-ed to mpz_t. The struct is,

```
typedef struct
{
    int _mp_alloc;
    int _mp_size;
    mp_limb_t *_mp_d;
} __mpz_struct;
```

where mp_limb_t points to a "limb" or digit. Its actual size depends on the system the code is compiled for but is typically an unsigned long int which is 4 bytes on the 32-bit system we assume in this book. The _mp_alloc field counts the number of limbs allocated while _mp_size is the number of limbs used. As we did above, the sign of the number is stored in the sign of _mp_size. If _mp_size is zero the number is zero. This saves allocating a limb and setting it to zero.

Input of big integers is via strings as was done in our code above. Output is to file streams or strings depending upon the function used. Integers must be created and initialized before being assigned but may be freely assigned after that. When done, they must be cleared. This code initializes a big integer, assigns it a value and then prints the value to stdout in two different number bases,

```
 1 |#include <stdio.h>
 2 |#include <gmp.h>
 3 |
 4 |int main() {
 5 |    mpz_t A;
 6 |
 7 |    mpz_init(A);
 8 |
 9 |    mpz_set_str(A, "1234567890123456789", 0);
10 |    mpz_out_str(NULL, 10, A);
11 |    printf("\n");
12 |    mpz_out_str(NULL, 2, A);
13 |    printf("\n");
14 |
15 |    mpz_clear(A);
16 |}
```

The GMP library is included in line 2 (linking with -lgmp is also required). A new big integer, A, is defined in line 5 and initialized in line 7. Note that A is not passed as an address. The definition of mpz_t is a one element array of _mp_struct which passes it as an address already. In line 9 we assign A from a string, the 0 means use the string itself to determine the base, in this case base 10. Lines 10 and 12 output A first as a base 10 number and then in binary. Finally, we release A in line 15.

Basic big integer functions in *mpz* include the expected addition (mpz_add), subtraction (mpz_sub) and multiply (mpz_mul). Also included are negation (mpz_neg) and absolute value (mpz_abs). For efficiency there are also special functions that combine operations as it is sometimes better to combine them than perform them individually. For example, there are mpz_addmul which adds the product of two numbers to an existing number as well as mpz_submul which subtracts the product.

For division *mpz* gives us many options. These options reflect various rounding methods and implement different algorithms which may be much faster than general algorithms if certain conditions are met. If we select only the tdiv functions, which round towards zero, like C does, we have three functions,

mpz_tdiv_q - divide n by d returning only the quotient, q.
mpz_tdiv_r - divide n by d returning only the remainder, r.
mpz_tdiv_qr - divide n by d returning both quotient, q, and remainder, r.

There is also a general mod function which always returns a positive result (mpz_mod). If it is known in advance that the numerator n is exactly divisible by the denominator d one may use mpz_divexact which is much faster than the generic division routines. Filling out the division options are mpz_divisible to check if n is exactly divisible by d and mpz_congruent_p to check if n is congruent to c modulo d meaning there is an integer q such that $n = c + qd$.

What algorithms are implemented by these top-level functions? For addition and subtraction, which are at the lowest level for Intel architecture implemented in assembly, we can expect the normal school method. What about multiplication? As we are now well aware, there are many options.

The GMP library uses one of seven multiplication algorithms for N digit by N digit multiplication (as of GMP version 6.0.0). Which is used depends upon the size of N and whether or not it exceeds a preset threshold. The algorithms are, in order of use (ie, larger N uses algorithms further down the list),

1. School method. This is the algorithm we implemented above.
2. Karatsuba. The Karatsuba method discussed above.
3. Toom-3. A three-way Toom extension to Karatsuba. See [7].
4. Toom-4. A four-way Toom algorithm.
5. Toom-6.5. A six-way Toom-like algorithm.
6. Toom-8.5. An eight-way Toom-like algorithm.
7. FFT. The Schönhage and Strassen method as discussed above.

All of these algorithms are optimized by the authors of GMP and at a lowest level are somewhat different than might be found in the original papers. If the operands to the multiplication are of very different sizes, N digits by M digits, then below one threshold the school method is used while FFT is used for very large numbers. In between, the Toom algorithms are used based on heuristics given the sizes N and M.

For division, four main algorithms are in use. For numbers below a threshold (50 digits) the long division method of Knuth is used. Above this threshold the divide-and-conquer method of Burnikel and Ziegler (modified) is used. Division by very large numbers uses an algorithm based on Barrett reduction [10]. Exact division uses Jebelean's method [11]. Switching between methods is automatic, the user need only call the top-level functions.

If GMP is for C and C++, what about other languages that lack intrinsic big integer functionality? Java uses the `BigInteger` class. This class comes with its own implementations, in Java, for low-level operations which exactly match those found in GMP's *mpn* package. This was done to make it possible to easily replace the Java code with highly performance C code if desired. So, we see that C, C++ and Java all make use of the GMP library.

For JavaScript a common choice to add support for big integers is the library `BigInteger.js` which is distributed via the MIT License and implements all functionality in pure JavaScript so it can run within a web browser. This implementation is very generic and uses a base of $10,000,000$ to take advantage of the simplicity this provides input and output just as we did with our implementation above. Addition and subtraction use the school method as does multiplication. There is a comment in version 0.9 to add Karatsuba in the future. Division uses the Knuth algorithm.

Languages. We know that Python has intrinsic support for big integers. Let's take a look at what it is doing under the hood. The Python interpreter is written in plain C and big integers are found in the source code file (Python 2.7) `longobject.c`. This file contains implementations for all the operations supported by Python long integers, it does not use the GMP or other libraries. For addition and subtraction the expected school method is used. For multiplication only two algorithms are available, the school method and Karatsuba if the number of digits exceeds 70. Division is long division *ala* Knuth. The numbers themselves use a base of 2^{15}. While the native implementations are adequate, for serious work in Python with big integers one would do well to wrap the GMP library. Surprisingly, given the current popularity of Python, this does not appear to have been done with any level of seriousness.

Another language which requires in its very definition the support of big integers is Scheme [12]. This is a small dialect of Lisp and has been used widely for instruction in computer programming. Indeed, it is used in the very popular *Structure and Interpretation of Computer Programs* by Abelson and Sussman [13] which has trained a generation of programmers from MIT though it has fallen out of style in recent years. Regardless, the text comes highly recommended.

The version of Scheme we will consider is *Scheme48* available online from *s48.org*. This version of Scheme conforms to the R5RS version of the language [14] and includes big integer support for exact numeric computation. Section 6.2.3 of the R5RS report ("Implementation Restrictions") strongly encourages implementations to support "exact integers and exact rationals of practically unlimited size and precision" which *Scheme48* does.

The Scheme interpreter itself is written in C and places big integer operations in the file bignum.c. Looking at this file we see that the implementations are based on earlier work from MIT. As expected, the addition and subtraction routines use the school method. The multiplication routine is limited to the school method as well, regardless of the size of the input. This limits the usefulness of Scheme for high-performance work with big integers unless using an external library like GMP. Similarly, we see that division is the Knuth algorithm, again regardless of the size of the operands.

4.6 Rational Arithmetic with Big Integers

Once we have the ability to represent any integer p with an arbitrary number of digits it logically follows that one might consider how to represent fractions, which are the ratio of two integers, using big integers. In this section we look at rational arithmetic with big integer components. If we group two big integers together, p and q, we can treat them as the numerator and denominator of a fraction, p/q, and perform arithmetic with them. For this section we will take advantage of the big integers provided by Python and extend them to implement basic rational arithmetic of arbitrary precision. It is proper to consider rational numbers in computers in a chapter on big integers because rational arithmetic involves fractions that grow quickly to have large numerators and denominators which implies that they will quickly exceed the capacity of fixed-length integers.

Elementary mathematics tell us how to implement the four basic arithmetic operations using fractions. They are,

$$\textit{Addition} \qquad \frac{a}{b} + \frac{c}{d} = \frac{ad+bc}{bd}$$

$$\textit{Subtraction} \qquad \frac{a}{b} - \frac{c}{d} = \frac{ad-bc}{bd}$$

$$\textit{Multiplication} \quad \frac{a}{b} \times \frac{c}{d} = \frac{ac}{bd}$$

$$\textit{Division} \qquad \frac{a}{b} \div \frac{c}{d} = \frac{ad}{bc}$$

where a,b,c, and d are themselves big integers. However, there is one twist we must bear in mind. If we were to implement these operations directly we would quickly find our numerators and denominators are becoming very large indeed. This is because after every operation we need to reduce the resulting fraction to lowest terms, as no doubt our grade school teachers mentioned to us many times! In order to do this we need to find, for each numerator and denominator, the greatest common divisor or GCD. Thankfully, this is straightforward to find as we will soon see.

Storing rationals in Python. Python is a modern object-oriented language. To store a rational number we need to store two integers, the numerator and the denominators. Since we are developing a rational arithmetic class, we will store the numerator and denominator as member variables n and d. Additionally, we will overload the basic math operations so that we can work with rationals in a manner similar to other numbers. Input and output will be straightforward as we simply display the number by printing the numerator, the fraction symbol (/), and the denominator. We will adopt the convention that the sign of the rational is stored in the numerator.

A basic rational arithmetic class. The first part of our `Rational` class defines the class and the constructor, `__init__`. Since Python uses automatic garbage collection we do not need in this case to define a specific destructor. So far, then, our class looks like this,

```
 1  from types import *
 2
 3  class Rational:
 4
 5      def __init__(self, n,d=None):
 6          if (d == None):
 7              t = float(n).as_integer_ratio()
 8              n = t[0]
 9              d = t[1]
10          g = self.__gcd(n,d)
11          self.n = n/g
12          self.d = d/g
13          if (self.d < 0):
14              self.n *= -1
15              self.d = abs(self.d)
```

where we consistently use four spaces for indenting as is the Python standard. In line 1 we include the `types` module which is part of the standard Python library. We will use this to determine whether or not arguments are `Rational` objects or integers.

Our constructor is in __init__ and takes self as the first argument as all Python methods do. This is a reference to the object itself and this is how we will access member data and call other class methods. The constructor takes two arguments, n and d, which are the numerator and denominator of the new rational object. Lines 6 through 9 check to see if a single argument was given. If so, it is assumed to be a floating-point number and converted to a rational by calling the as_integer_ratio method of the float class. In line 10 we call the __gcd method to calculate the GCD of the given numbers. We will explore this method in more detail shortly. Note that we are using the standard Python naming convention of two leading underscores (__) which marks the method as being private to our class. The important point here is that each method in our class representing an operation will return a new Rational object. This means that the only time we need consider reducing the fraction is when the new object is created as it is not changed once created. In lines 11 and 12 we set the local values for the numerator (self.n) and denominator (self.d) dividing each by the GCD so that the fraction is stored in lowest terms. For simplicity, we ignore error states such as a denominator of zero. Lines 13, 14 and 15 ensure that the new fraction places its sign in the numerator as is our convention. As this point, we have a new Rational object and it is initialized with the given numerator and denominator in lowest terms. Now lets look at the __gcd method itself.

Reducing fractions using the GCD. In school we learned that fractions are not unique. For example, p/q is a fraction with a particular value which we can say, when expressed as a decimal number, is c since $c = p/q$. But, if we multiply our fraction by a new fraction which is equal to 1, $\alpha/\alpha = 1$, we get $\alpha p/\alpha q$ which is, of course, still c. Therefore, when working with fractions we were constantly encouraged to express our answers in lowest terms (factor out the α). When working with rational numbers on a computer this is especially important to prevent our numerators and denominators from growing too large too quickly. To reduce the fraction we need to find α which is the greatest common divisor of both p and q and divide p and q by it.

The iterative form of the algorithm of Euclid [15] is a good way to calculate the GCD even for large numbers. Our implementation of it is,

```
1 def __gcd(self, a,b):
2     while b:
3         a,b = b, a % b
4     return abs(a)
```

Arithmetic and other methods. We are now ready to implement basic arithmetic operations. Python defines a set of methods that users can override in their own classes to capture situations which syntactically look like more ordinary operations. For example, the seemingly simple statement,

```
a + b
```

is really equivalent to,

```
a.__add__(b)
```

which means that a Python class which overrides the __add__ method will be able to participate in mathematical expressions. This is exactly what we require in order to make using the Rational class as natural as possible.

Python has a large set of methods that we could override in our class. We will do some here and leave others to the exercises. In particular, we will override the following methods,

__add__	Addition (+)
__sub__	Subtraction (−)
__mul__	Multiplication (×)
__div__	Division (/)
__str__	String output for str() and print
__repr__	String output for repr()
__float__	Floating-point value for float()
__int__	Integer value for int()
__long__	Integer value for long()

In addition, we will override the "reverse" forms for the arithmetic operators. These are called if instead of a+2 we write 2+a. For operators that commute, addition and multiplication, we really only need call the non-reversed form. For division and subtraction we have to be more careful.

So, how do we implement addition? Like so,

```
 1 def __add__(self, b):
 2     t = type(b)
 3     if (t == InstanceType):
 4         n = self.n*b.d + self.d*b.n
 5         d = self.d*b.d
 6     elif (t == IntType) or (t == LongType):
 7         n = self.n + self.d*b
 8         d = self.d
 9     return Rational(n,d)
10
11 def __radd__(self, b):
12     return self.__add__(b)
```

where we first look at the type of the argument, b, in line 2. We do not only want to work with other rational numbers but also with integers and long integers. If the argument is another rational number we move to line 4 and add to get the numerator followed by the denominator. Otherwise, if the argument is an integer, we move to line 7 to add the integer to the fraction. Line 9 returns the new rational answer. Recall that the constructor is what applies the GCD so it is not explicitly called here. Line 11 defines the reverse case method. Since addition is commutative, $a + b = b + a$, we simply fall back to the main addition method.

Subtraction is virtually identical except for the reversed call,

```
 1 def __sub__(self, b):
 2     t = type(b)
 3     if (t == InstanceType):
 4         n = self.n*b.d - self.d*b.n
 5         d = self.d*b.d
 6     elif (t == IntType) or (t == LongType):
 7         n = self.n - self.d*b
 8         d = self.d
 9     return Rational(n,d)
10
11 def __rsub__(self, b):
12     t = type(b)
13     if (t == IntType) or (t == LongType):
14         n = b*self.d - self.n
15         d = self.d
16         return Rational(n,d)
```

where __sub__ is identical to __add__ except for the change from + to −. However, as subtraction is not commutative, we must implement it explicitly for the reverse case. This is done in lines 13 through 16. Why is there no check for InstanceType? Because Python is smart enough to know that if both arguments are objects the main method will be called instead.

Multiplication follows addition,

```
 1 def __mul__(self, b):
 2     t = type(b)
 3     if (t == InstanceType):
 4         n = self.n * b.n
 5         d = self.d * b.d
 6     elif (t == IntType) or (t == LongType):
 7         n = self.n * b
 8         d = self.d
 9     return Rational(n,d)
10
11 def __rmul__(self, b):
12     return self.__mul__(b)
```

while division follows subtraction in implementing the reverse call explicitly,

```
 1 def __div__(self, b):
 2     t = type(b)
 3     if (t == InstanceType):
 4         n = self.n * b.d
 5         d = self.d * b.n
 6     elif (t == IntType) or (t == LongType):
 7         n = self.n
 8         d = self.d * b
 9     return Rational(n,d)
10
11 def __rdiv__(self, b):
12     if (type(b) == IntType):
13         n = b * self.d
14         d = self.n
15         return Rational(n,d)
```

and in both cases the proper steps for multiplication and division of fractions are performed.

The remaining methods are convenience methods,

```
1  def __str__(self):
2      if (self.n == 0):
3          return "0"
4      elif (self.d == 1):
5          return "%d" % self.n
6      else:
7          return "%d/%d" % (self.n, self.d)
8
9  def __repr__(self):
10     return self.__str__()
11
12 def __float__(self):
13     return float(self.n) / float(self.d)
14
15 def __int__(self):
16     return int(self.__float__())
17
18 def __long__(self):
19     return long(self.__float__())
```

which allow us to interact with Python in a more natural way. Use __str__ to return a string representation of the fraction as the output of str() or print. In this method we check for special cases like zero or an integer (denominator of one). The __repr__ method responds to the repr() function by returning exactly the same string print would receive. The __float__, __int__ and __long__ methods perform the necessary conversion to return the rational as a float, integer, or long integer, respectively. These are used by the float(), int(), and long() functions. These are the minimal set of methods for our Rational class.

Using the Rational class. Because of our design choices in overriding key default Python methods we can use the Rational class in a quite natural way. The following examples are from an interactive session and illustrate how to use the class,

```
1  >>> from Rational import *
2  >>> a = Rational(2,3)
3  >>> b = Rational(355,113)
4  >>> c = a + b
5  >>> c
6  1291/339
7  >>> (a+b) * c
8  1666681/114921
9  >>> d = (a + b) * c
10 >>> d = 2*d
11 >>> d
12 3333362/114921
13 >>> float(b)
14 3.1415929203539825
15 >>> d / d
16 1
17 >>> 1/d
18 114921/3333362
19 >>>
```

where >>> is the Python interactive shell prompt and replies from Python are lines without the prompt.

In line 1 we import the Rational class. Lines 2 and 3 define a and b to be 2/3 and 355/113, respectively. Line 4 adds a and b and assigns the result to c which is then displayed in line 6. Line 9 uses all three rationals in an expression and displays it. Note already how the numerator and denominator are growing. We define d in line 9 and multiply it by an integer in line 10. Line 13 shows us that 355/113 is a good approximation of π by requesting the floating point version of it. We use line 15 as a sanity check to make sure that a rational number acts rationally. Finally, line 17 checks that the inverse of a rational is, in fact, what you get when you flip the numerator and the denominator.

This concludes our simple Python Rational class. As promised, the exercises will offer you the opportunity to extend the class even further.

4.7 When to Use Big Integers and Rational Arithmetic

The sections above demonstrate how to implement big integers, how to operate on them efficiently, and where to find performant big integer libraries, however, none of the above actually talks about *when* and *why* one might want to use big integers. We remedy that in this section.

In modern computing undoubtedly the largest use of big integers is in cryptography. Cryptographic systems depend upon the fact that certain operations are easy to do in one direction but next to impossible to do, in a reasonable period of time, in the other direction. Most of these systems depend upon the following two facts,

1. The *fundamental theorem of arithmetic* states that all numbers are either prime or can be factored, uniquely, into a specific set of primes.
2. Factoring large integers is a tedious process even though multiplying them is straightforward.

This means that while it is easy to find $n = pq$ with p and q both integers it is hard to factor n to see that it is in fact what you get if you multiply p and q. This is the second fact above. The first says that if p and q are prime then indeed the *only* way to factor n is to find the specific pair, p and q. If p and q are too small, trial and error will eventually find them, but if they are very large, on the order of several hundred bits or about 100 decimal digits, then finding the factor of their product will take a very long time. This is the heart of modern cryptography. Let's see briefly how multiplying large integers plays into it.

First we consider classic Diffie-Hellman key exchange [16]. This system allows two parties, traditionally known as Alice and Bob, to exchange a secret number to be used as a key to encrypt a message shared between them. We are not concerned with the encryption but rather how they exchange the secret key so that an outsider (traditionally known as Carol or Eve) cannot learn what it is. Let's look at the steps and then we will see where big integers play into it,

The heart of the exchange relies on the choice of p and g. We must choose p to be prime and g to be a primitive root modulo p. The exact definition of a "primitive root modulo p" does not concern us here but it a number theory concept. What does concern us is that p and g are public and known by Alice, Bob and Carol. Additionally, Alice and Bob each select a secret number only they know. We'll call Alice's number a and Bob's number b. These numbers are known only to the respective person. For our example, we will select $p = 37$ and $g = 13$, which is a primitive root of 37. With this setup, the steps necessary to exchange a secret key are,

1. Alice chooses $a = 4$. Only Alice knows this number.
2. Bob chooses $b = 11$. Only Bob knows this number.
3. Alice sends Bob $A = 13^4 \bmod 37 = 34$. Carol sees A.
4. Bob sends Alice $B = 13^{11} \bmod 37 = 15$. Carol sees B.
5. Alice computes the secret key, $s = 15^4 \bmod 37 = 9$. Carol cannot compute s.
6. Bob computes the secret key, $s = 34^{11} \bmod 37 = 9$.
7. Alice and Bob both use s as the key for their encryption system.

Recall that a and b are secret but $p = 37$, $g = 13$, $A = 34$ and $B = 15$ are public. When the exchange is complete, Alice and Bob have both calculated $s = (g^a)^b \bmod 37 = (g^b)^a \bmod 37 = 9$ which they will use as the secret key for their encryption system without Carol knowing what s is.

Big integers come into this system with the choice of p, a and b. If these are very large, hundreds of digits long instead of small as in our example above, it becomes virtually impossible to determine s from the values that are known publicly. Even though A is known and it is known that $A = g^a \bmod p$, the reverse operation, known as the *discrete logarithm*, is very hard to do so a remains unknown publicly. It is this difficulty that makes the key exchange system secure and big integers are essential because without them the numbers involved would be too small to offer security against brute force attacks.

Another cryptographic system that relies on big integers is RSA public key encryption [17]. This system was published by Rivest, Shamir, and Adelman (hence RSA) of MIT in 1977. In this system users generate two keys, a public key and a private key. They then publish their public key allowing anyone to encrypt a message using it. The holder of the corresponding private key, however, is the only one who can decrypt it. For simplicity we ignore the extra twist of signing the message using the sender's private key. The essence of the algorithm depends upon two primes, p and q, their product, $n = pq$, a public key exponent, e, and its multiplicative inverse, d. The person generating the keys keeps d as the private key and releases both n and e together to be the public key. With this public key anyone can encrypt a message but only the holder of the private key as decrypt it. The length of n is the key length which is related to the strength of the algorithm.

If Bob has released his public key, the pair n and e, then Alice can send Bob a secure message by turning her message into an integer, m, the exact method for this is irrelevant but must be agreed upon by both Alice and Bob, and computing $c = m^e \bmod n$. She then sends c to Bob. Bob, in turn, recovers Alice's message by computing $m = c^d \bmod n$ using the private key, d. Since only Bob has d, he is the only one who can recover the message.

As in Diffie-Hellman, the security of the encryption depends upon the choice of large primes for p and q. When these are very large numbers it becomes virtually impossible to recover them from knowledge of just n. The exponent, e, and private key, d are calculated from p and q so knowledge of p and q would lead to a breakdown of the entire algorithm. Fortunately, big integers come to our rescue and let us work with the very large primes that make the system secure. Practical implementations of encryption algorithms which rely on big integers are found in *OpenSSL* [18] which implements the secure socket layer for internet data transfer and *libgcrypt* [19] which is an open source package implementing numerous algorithms including RSA.

Big integers are of practical importance to cryptography but they are also used for research purposes by mathematicians. For example, the open source package *Pari* [20] has been in existence for some time and it supports arbitrary precision calculations for mathematicians interested in number theory and other disciplines. Additionally, sometimes it is helpful to have the ability to restrict the precision of calculations in order to explore the effect of mathematical operations. In the paper *Chaos, Number Theory and Computers* [21] Adler, Kneusel and Younger explored the effect of the precision used when calculating values for the logistic map, $x_{n+1} = Ax_n(1-x_n)$, $A \leq 4$, $0 < x < 1$, which can lead to anomalous behavior. The software for this paper made use of the intrinsic big integer arithmetic found in the Scheme language to calculate logistic map values with a set precision.

What about rational arithmetic? As we've seen, adding rational arithmetic once big integers are available is straightforward. Many languages and libraries do support rational arithmetic. For example, the mpq part of GMP implements rational functions and while Python supports big integers natively, Scheme supports rationals natively as well including syntactically so that (+ 2/3 3/4) is a perfectly valid function call yielding 17/12 as its answer. Another use for rational arithmetic would be as a way to increase the precision of calculations beyond that of native floating point.

4.8 Chapter Summary

In this chapter we introduced the concept of a big integer by which we mean an integer larger than any native type supported by the computer. We implemented basic big integer operations in C and discussed improved algorithms for these basic operations which are of more practical utility. Next we looked at important implementations of big integer libraries as well as programming languages which support big integers intrinsically. We then examined rational numbers built from big integers and implemented a simple rational number class in Python. Lastly, we discussed some prominent uses of big integers to give examples of when big integers might be the tool of choice.

Exercises

4.1. Big integer libraries typically represent integers in *sign-magnitude* format as we did in this chapter. Another possibility would be to use a complement notation, call it B's-complement where B is the base, which is the analog of 2's-complement used almost exclusively for fixed-length integers. However, very few big integer packages actually do this. Why?

4.2. In Sect. 4.2 we gave a C function for bigint_equal to test if two big integers are equal. This function made use of bigint_compare. Write the remaining comparison functions for $<, >, \leq, \geq$ and $\neq$. *

4.3. Figure 4.11 implements a function to perform big integer division. Extend this function to return the remainder as well. *

4.4. Add a new function to our library, bigint_pow(a,b), which returns a^b. Be sure to handle signs and special cases. *

4.5. Add another new function to our library, bigint_fact(a), which returns $a!$. Recall that the argument must be ≥ 0. *

4.6. Using the big integer functions defined in this chapter create a simple arbitrary precision desk calculator in C which takes as input expressions of the form "*A op B*", as a string, where A and B are big integers and *op* is one of the four basic arithmetic operations and returns the result. If you wish, you may use postfix notation and a stack. In postfix notation the operands come first followed by the operation, "*A B op*". You can use Python to check your answers. **

4.7. Extend the basic Rational Python class outlined above to compare rational numbers by overriding the __lt__, __le__, __gt__, __ge__, __eq__, and __ne__ methods to implement comparison operators $<, \leq, >, \geq, ==$, and $\neq$. *

4.8. Extend the basic Rational Python class outlined above by adding __neg__, __abs__ methods for negation and absolute value. *

4.9. Extend the basic Rational Python class outlined above by adding methods to handle shortcut operations. N.B., these, unlike all the others, destructively update the existing rational object. They do not return a new object. The method names are: __iadd__, __isub__, __imul__, and __idiv__ corresponding to operators +=, -=, *=, and /=. Be careful when handling integer arguments and be aware that the updated value may no longer be in lowest terms. *

4.10. Use the Rational class to create a simple interactive rational desk calculator. This should take as input expressions of the form "*A op B*", as a string, where A and B are rationals (with a "/" character) or integers and *op* is one of the four basic arithmetic operations and return the result. If you wish, you may use postfix notation and a stack. **

References

1. Crandall R., Pomerance C., Prime Numbers: a Computational Perspective, Springer (2005).
2. Comba P., Exponentiation Cryptosystems on the IBM PC. IBM Systems Journal 29, 4:526–536 (1990).
3. Karatsuba, A., Ofman, Y., Multiplication of Many-Digital Numbers by Automatic Computers, Proceedings of the USSR Academy of Sciences 145: 293–294 (1962). Translation in the academic journal Physics-Doklady, 7 , pp. 595–596 (1963).
4. Schönhage, D., Strassen, V., Schnelle multiplikation grosser zahlen, Computing, 7(3–4), 281–292 (1971).
5. Saracino D., Abstract Algerbra: A First Course, Addison-Wesley (1980).
6. Bracewell, R., The Fourier transform and its applications, New York: McGraw-Hill (1986).
7. Knuth, D., The Art of Computer Programming, Volume 2: Seminumerical Algorithms, Addison-Wesley Professional (2014).
8. Burnikel, C., Ziegler, J., Im Stadtwald, D., Fast recursive division (1998).
9. Granlund, T., gmp – GNU multiprecision library, Version 6.0.0 (2014).
10. Barrett, P., Implementing the Rivest Shamir and Adleman Public Key Encryption Algorithm on a Standard Digital Signal Processor, Advances in Cryptology – CRYPTO' 86. Lecture Notes in Computer Science 263. pp. 311–323 (2006).
11. Jebelean T., An algorithm for exact division, Journal of Symbolic Computation, volume 15, 169–180 (1993).
12. Sussman, G. Steele, Jr, G., The First Report on Scheme Revisited, Higher-Order and Symbolic Computation 11 (4): 399–404 (1998).
13. Abelson, H., Sussman, G., Structure and interpretation of computer programs, 2nd Ed, MIT Press (1996).
14. Kelsey, R., Clinger, W., Rees, J. (eds.), Revised5 Report on the Algorithmic Language Scheme, Higher-Order and Symbolic Computation, Vol. 11, No. 1, (1998).
15. Heath T., The Thirteen Books of Euclid's Elements, 2nd ed. (Facsimile. Original publication: Cambridge University Press, 1925) , Dover Publications (1956).
16. Diffie, W., Hellman, M., New directions in cryptography, Information Theory, IEEE Transactions on, 22(6), 644–654 (1976).
17. Rivest, R., Shamir, A., Adleman, L., A method for obtaining digital signatures and public-key cryptosystems, Communications of the ACM, 21(2), 120–126 (1978).
18. http://www.openssl.org/.
19. http://www.gnu.org/software/libgcrypt/.
20. The PARI Group, PARI/GP version 2 . 7 . 0, Bordeaux, 2014, http://pari.math.u-bordeaux.fr/.
21. Adler C., Kneusel R., Younger, B., Chaos, Number Theory, and Computers, Journal of Computational Physics, 166, 165–172 (2001).

Chapter 5
Fixed-Point Numbers

Abstract Fixed-point numbers implement floating-point operations using ordinary integers. This is accomplished through clever use of scaling and allows systems without explicit floating-point hardware to work with floating-point values effectively. In this chapter we explore how to define and store fixed-point numbers, how to perform signed arithmetic with fixed-point numbers, how to implement common trigonometric and transcendental functions using fixed-point numbers, and, lastly, discuss when one might wish to use fixed-point numbers in place of floating-point numbers.

5.1 Representation (Q Notation)

A fixed-point number is a scaled floating-point number. The scaling can be in decimal or binary, we will focus on binary here because this makes the scaling a power of two which means scaling up or down is simply a matter of shifting bits up or down. One reason to use fixed-point arithmetic is speed, so this rapid shifting is an advantage. Let's take a look at what scaled arithmetic actually entails.

If we are working with numbers in a known range, say between -3 and $+3$, we normally use a `float` or `double` to represent the numbers and go on our merry way. However, what if we are in a situation where floating-point math is expensive in terms of speed or memory? One option is to use scaled arithmetic, which is fixed-point arithmetic. Before we can implement any arithmetic functions, however, we need to understand the way scaling is used so let's look at an example. If we have 16-bit integers, C `short int`, we can represent values between -65536 and 65535, including the end points, if we use two's-complement notation. This means that we need to scale our -3 to $+3$ range numbers so that they fit in the range -65536 to 65535. Additionally, we want to make the scale factor a power of two and make it as large as possible while still keeping our scaled integer in the 16-bit range. In this case, we need two bits to store the integer part and one additional bit to store the sign for a total of three bits. This leaves $16 - (2 + 1) = 13$ bits for

© Springer International Publishing Switzerland 2015
R.T. Kneusel, *Numbers and Computers*, DOI 10.1007/978-3-319-17260-6_5

representing the fractional part. This means that the scale factor is $2^{13} = 8192$ and the resolution of our fixed-point numbers will be 2^{-13} or about 0.000122. Here "resolution" means that the difference between any two consecutive 16-bit integers, in this interpretation, is a fixed difference of 0.000122. More formally, to implement scaled arithmetic, we are really using integers to represent rational numbers with an implied denominator which is the scale factor. So, for our example, we are using 16-bit integers to represent fractions where the integer is the numerator and the denominator is fixed at $2^{13} = 8192$. We will look at how to convert floating-point numbers to and from fixed-point notation and then things should be clearer but before doing that, let's introduce some notation that will make discussing fixed-point numbers much easier, this is Q notation.

The Q notation was introduced by Texas Instruments (for example, see [1]) as a shorthand for specifying the size of the integer used to represent a fixed-point number as well as the number of bits reserved for the integer and fractional part. Formally, the notation is $Qm.n$ where m is the number of bits used in the integer part while n is the number of bits in the fractional part. Signs are stored using two's-complement notation meaning the entire integer is a two's-complement number. As this is the expected way fixed-length integers are stored in modern computers this is a good thing. This also means that there must be one additional bit for the sign of the fixed-point number and the entire length of the fixed-point number is $m + n + 1$ bits. Our example above used two bits for the integer part and 13 for the fractional part. In Q notation this is $Q2.13$ which uses $2 + 13 + 1 = 16$ bits, exactly the number of bits in a C `signed short` integer. If the number of integer bits is zero it is dropped from the Q notation so that $Q31$ means a 32-bit integer with no integer part and 31 bits reserved for the fractional part.

Armed with Q notation we can see that our example above is a good choice for representing numbers in the range -3 to $+3$ using 16 bits. The range of a fixed-point number $Qm.n$ is $[-2^m, 2^m - 2^{-n}]$ which for $Q2.13$ is about $[-4, 3.999878]$, beyond the $[-3, 3]$ range we required. By picking the smallest number of bits for the integer part we maximize the number of bits for the fractional part thereby making the fixed-point numbers as accurate as possible over the desired range for the given total number of bits.

What if instead of selecting $Q2.13$ we select $Q7.8$? In this case the interval between values is 2^{-8} or about 0.0040 which is 32x larger than $Q2.13$ meaning our numbers are much less accurate. To see this, consider a number line such as that of Fig. 5.1 which shows graphically the reduced accuracy we would have if we did not keep the integer part of the fixed-point number as close to the working range as possible.

We now understand that a fixed-point number is a fraction with a denominator that is a power of two. We also know how to indicate, via the Q notation, how we are interpreting this fraction so that the entire range of the fixed-width integer we are using to represent the fraction is used. With this in mind, we can begin to piece together our own library of fixed-point functions in C. In this case C is a very natural choice as one of the primary uses of fixed-point is in embedded development where often one is working with microcontrollers without floating-point hardware.

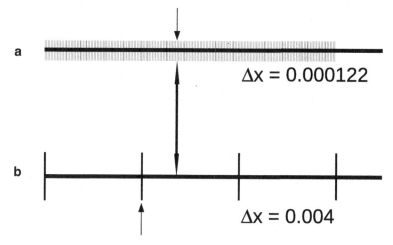

Fig. 5.1 The difference between numbers representable in (**a**) $Q2.13$ and (**b**) $Q7.8$. For $Q2.13$ each hash mark is 0.000122 apart while for $Q7.8$ the hash marks are 0.0040 apart. If we want to represent a floating point value, indicated by the double arrow, we must select the closest hash mark in each subfigure (*small arrows*). This is why if the range is known we should select the smallest integer width encompassing that range

While it is possible, and often done in the past, to emulate floating-point hardware in software, in general, fixed-point math is much faster. And in the microcontroller world, C is still the language of choice.

Existing fixed-point libraries are highly optimized to take full advantage of the performance found in small systems. To this end, they often implement one or two types of fixed-point numbers. For example, the open source *libfixmath* project [2] implements $Q15.16$ only using signed 32-bit integers. This is a reasonable choice for a general-purpose library, actually, as it covers the range $[-32768, 32767.99998]$ which likely includes many of the values one would encounter in daily practice, especially in the embedded development world. The resolution is 2^{-16} or about 0.0000153 which is also quite reasonable. For our library, however, we will opt to implement a variable set of fixed-point numbers. This means that when the library is initialized we will pass in our desired $Qm.n$ by passing m explicitly. This will determine how we will interpret values internally and allow us to tailor the library to maximize the floating-point accuracy of our numbers. The library will always use a signed 32-bit integer to represent the number. This means that we need only specify m, the number of bits for the integer part, and we know that n will be $32 - (m + 1)$ bits so that the total number of bits used is always 32 (recall one extra bit for the sign). Additionally, when the library is initialized for a particular configuration we will recalculate certain constants which will be used when we implement trigonometric and transcendental functions.

We start the library with code to set up for a particular $Qm.n$ and to calculate some constants in that representation for future use. Note that we create a new type, fix_t, to distinguish between ordinary integers and integers we mean to interpret as fixed-point numbers as well as $fixx\_t$ for a double precision 64-bit version. So, our initialization routine becomes,

```
1  #define MASK_SIGN 0x80000000
2  #define DEG2RAD 0.0174532925
3  #define PI 3.141592653589793
4  #define EXP_ITER_MAX 50
5  #define EXP_TOLERANCE 0.000001
6
7  typedef int32_t fix_t;
8  typedef int64_t fixx_t;
9  int32_t M=0,N=31;
10 fix_t PIh, PId, d2r, r2d;
11 fix_t f90,f180,f270,f360;
12 fix_t st1,st2,st3,st4;
13 fix_t p0,p1,p2,p3,p4,p5;
14 fix_t sin_tbl[91];  // sine lookup table
15
16 void Q_init(int32_t m) {
17     int i;
18
19     M = ((m >= 0) && (m < 32)) ? m : 0;
20     N = 32 - (m+1);
21
22     for(i=0; i<91; i++)
23         sin_tbl[i] = Q_to_fixed(sin(i*DEG2RAD));
24
25     PId = Q_to_fixed(2*PI);
26     PIh = Q_to_fixed(PI/2);
27     d2r = Q_to_fixed(DEG2RAD);
28     r2d = Q_to_fixed(1.0/DEG2RAD);
29     f90 = Q_to_fixed(90*DEG2RAD);
30     f180= Q_to_fixed(180*DEG2RAD);
31     f270= Q_to_fixed(270*DEG2RAD);
32     f360= Q_to_fixed(360*DEG2RAD);
33     st1 = Q_to_fixed(1.0/6);
34     st2 = Q_to_fixed(1.0/120);
35     st3 = Q_to_fixed(1.0/5040);
36     st4 = Q_to_fixed(1.0/362880);
37     p0  = Q_to_fixed(1.17708643e-05);
38     p1  = Q_to_fixed(9.99655930e-01);
39     p2  = Q_to_fixed(2.19822760e-03);
40     p3  = Q_to_fixed(-1.72068031e-01);
41     p4  = Q_to_fixed(5.87493727e-03);
42     p5  = Q_to_fixed(5.79939713e-03);
43 }
```

where Q_init takes the number of integer bits in m. We check that m is reasonable in line 19 and set our global M (line 9) to it. In line 20 we set N so that we always use 32 bits for the entire number. Note that we are including two standard libraries (not listed). The *math.h* library is necessary for our conversion routines between floating-point and fixed-point and to be explicit about integer types we include *stdint.h*.

This latter library defines the integer types in a consistent way so that `int32_t` is a `signed int` of 32-bits while `uint32_t` is an `unsigned int` of 32-bits.

We introduce several `#define` values. We use `MASK_SIGN` to mask values to get the sign of the number. This will be handy when we check for overflow. `DEG2RAD` is the constant that converts an angle in degrees to radians. This is used in line 23 to build the sine look up table. We define `PI` to be π for convenience. We will use `EXP_ITER_MAX` and `EXP_TOLERANCE` when we implement the exponential function.

Besides M and N there are several other globals which in a fixed library would themselves be constants. Here they are variables so that we can adapt to different Q representations. In line 10 we define variables to hold fixed-point representations of 2π, $\pi/2$, `DEG2RAD`, and its reciprocal, respectively. The specific assignments for a new Q notation are in lines 25 through 28.

The variables `f90`, `f180`, `f270`, and `f360` are set to $\pi/2$, π, $3\pi/2$ and 2π radians. These values are needed when implementing the sine function. Variables `st1`, `st2`, `st3`, and `st4` are set to coefficients of the Taylor series expansion of $\sin x$. Lastly, `p0` through `p5` are the coefficients of the 5-th degree polynomial fit to $\sin x$. These values are determined empirically by curve fitting.

We will add two convenience functions to the library which one would likely not include if working in a constrained environment. These functions let us easily map a floating-point value (C `double`) to a fixed-point value and back. To find the $Qm.n$ representation of a floating-point value simply scale it by the denominator, 2^n, and round to the nearest integer,

$$z_{fixed} = \lfloor 2^n z_{float} + 0.5 \rfloor$$

Similarly, to change fixed-point back to floating-point simply multiply by the reciprocal of the denominator,

$$z_{float} = 2^{-n} z_{fixed}$$

These functions are easily implemented in C as,

```
1  fix_t Q_to_fixed(double z) {
2      return (fix_t)floor(pow(2.0,N)*z + 0.5);
3  }
4
5  double Q_to_double(fix_t z) {
6      return (double)z * pow(2.0,-N);
7  }
```

What about comparisons of fixed-point numbers? In this case, since we are scaling by a fixed power of two (ignoring the case of comparing different Q notations), normal integer comparisons will still work as expected so we do not need to implement anything specific for fixed-point numbers. An example will illustrate. If we use an eight bit integer and $Q3.4$ notation to store 2.6875 we interpret each bit as follows,

bit value →	0	0	1	0	1	0	1	1
place value →	$sign$	2^2	2^1	2^0	2^{-1}	2^{-2}	2^{-3}	2^{-4}

where the integer we store has the value $2b_{16} = 43$. If we then look at storing 2.3125 in the same format we get an integer with the value of 37. Clearly, since $2.3125 < 2.6875$ we also have $37 < 43$ following directly from the fact that if the Q notation is the same then the place value for each bit is the same and we can compare bit by bit as is done when comparing two signed integers. Negative values are stored in two's-complement format so we again see that the comparison of two signed integers arrives at the same relationship as the two floating-point values those integers represent when interpreted as fixed-point numbers. With this understood, we are now ready to look at implementing basic arithmetic.

5.2 Arithmetic with Fixed-Point Numbers

Arithmetic with fixed-point numbers is arithmetic using fractions with an implied denominator. Another way to consider fixed-point numbers is that they are scaled by the fractional part. So, if x is a fixed-point representation of X, a floating-point number, then $x = Xd$ where $d = 2^n$ is the scale factor, with rounding to the nearest integer. Similarly, for fixed-point number $y = Yd$, where Y is the actual floating-point number, we have,

$$x + y = Xd + Yd = (X + Y)d$$

and,

$$x - y = Xd - Yd = (X - Y)d$$

which means that addition and subtraction of fixed-point numbers naturally results in a properly scaled answer. If the magnitude of $X + Y$ or $X - Y$ is too large to fit in m bits then the resulting number will overflow or underflow the representation and a loss of precision, possibly significant, will occur.

For multiplication we have,

$$xy = (Xd)(Yd) = XYd^2$$

where the answer is now scaled by d^2 instead of d. To get back to the same scaling we divide by d to get a final answer, XYd, which is what we expect for multiplication of two $Qm.n$ numbers returning a $Qm.n$ result.

There are several things to keep in mind with multiplication of fixed-point numbers. First, multiplication may lead to an intermediate result that requires double the precision to hold since it is scaled by d^2 instead of d. Second, a loss of fractional precision may well happen when we divide by d to get the final answer. This loss will be in the lowest bits of the fractional part. Once the division has been made the resulting value can be put back into a single precision integer. This means keeping the lowest b bits of the double precision $2b$-bit integer. Note that the division is simply a shift to the right by n bits for a $Qm.n$ number. Finally, if the integer part of the multiplication has overflowed meaning the result has an integer value that cannot fit in m bits, there will be a catastrophic loss of precision in the answer. Therefore, it is important to check for this overflow when working with the double precision version.

There is one other detail we did not mention yet. This is the issue of what happens when we divide the double precision product by the denominator. Recall that this division is really a shift to the right a specified number of bits. If we do this shift without concern for any rounding issues we will introduce bias in our results. To avoid this, we take the extra step of adding a value that is equal to 2^{n-1} which is the highest bit position that will be discarded by the right shift of n bits. This is tantamount to adding 0.5 to a floating point value before applying the *floor* operation so that the resulting integer is the integer closest to the number being rounded. This will increase the precision of our answers in the long run and cause errors to average out over a series of calculations without constantly losing precision by applying a *floor* operation. For multiplication this extra addition is done before the rescaling by 2^{-n} so that we use the bits that will be discarded.

The algorithm for multiplication of two $Qm.n$ fixed-point numbers, x and y, is,

1. Cast x to x' which is a sign-extended $2b$-bit integer. Cast y to y', also a sign-extended $2b$-bit integer.
2. $t' = x' \times y'$. This is the result scaled by $(2^n)(2^n) = 2^{2n}$ and stored in a $2b$-bit integer, t'.
3. $t' = t' + 2^{n-1}$. This is the rounding to avoid truncation bias. We will divide by 2^n so we add 2^{n-1} which is the largest bit value we are about to discard.
4. $t' = t' \times 2^{-n}$. This is the result scaled by 2^n. Note that this is a right shift by n bits.
5. Cast t' to t by keeping the lowest $m+n+1$ bits. This is the product in $Qm.n$ format. If any bits of t' above bits $m+n+1$ are set overflow has happened and our result, t, will be incorrect.

Lastly, let's look at fixed-point division. We want to divide two $Qm.n$ numbers, x by y, and return a quotient that is also a $Qm.n$ number. This means we want an integer that represents X/Y scaled by 2^n. This is what we get if we implement the following algorithm,

1. Cast x to $x' = x \times 2^n$ where x' is a $2b$-bit integer. This means that x' is x scaled by 2^{2n}, hence using a $2b$-bit integer to store the result.
2. Correct for truncation bias by adding $y/2$ to x', $x' = x' + y/2$. Note that $y/2$ is also a $2b$-bit integer. This addition is adding $\frac{Y \times 2^n}{2} = Y \times 2^{n-1}$ to the value x' which is scaled by 2^{2n}. We will see below why this is helpful.
3. Calculate a quotient, $q' = x'/y$ where y has been properly sign-extended to $2b$-bits. Because of the initial scaling of x this quotient is already properly scaled so we keep the lowest b bits to return the $Qm.n$ answer.

Let's look more closely at this algorithm. In the end, we want a number that is $(X/Y) \times 2^n$ which is the scaled $Qm.n$ representation of the quotient of $x/y = X/Y$. By scaling x in step 1 we have $x' = x \times 2^n = X \times 2^{2n}$. The bias correction in step 2 seems particularly mysterious but we see that we are adding $y/2 = (Y \times 2^n)/2 = Y \times 2^{n-1}$. This means that the division in step 3 is,

$$x'/y = \frac{X \times 2^{2n} + Y \times 2^{n-1}}{Y \times 2^n} = (X/Y) \times 2^n + \frac{1}{2}$$

which is exactly the quotient we want, properly scaled to $Qm.n$ plus the bias correction of $1/2$.

Now that we know how to do basic arithmetic with fixed-point numbers, let's implement these operations for our library. We will implement two versions of addition, subtraction, and multiplication, one that checks for overflow and one that does not. First we present versions that do not check for overflow conditions as they are the simplest,

```
1  fix_t Q_add(fix_t x, fix_t y) { return x+y; }
2  fix_t Q_sub(fix_t x, fix_t y) { return x-y; }
3
4  fix_t Q_mul(fix_t x, fix_t y) {
5      fixx_t ans = (fixx_t)x * (fixx_t)y;
6
7      ans += 1<<(N-1);
8      return (fix_t)(ans >> N);
9  }
```

Addition and subtraction are as simple as can be if we do not check for overflow. Adding (subtracting) two two's-complement integers is mathematically identical to adding two "floating-point" values stored in two's-complement format which is what a fixed-point number actually is. There is an implied radix point between bits n and $n - 1$, counting bits from zero, but the process of addition (subtraction) of the fractional part is the same as for the integer part so we simply add or subtract the two values and will arrive at the proper answer.

Multiplication is much the same but if we think about the school method of multiplication we see that we will end with, since both numbers are in $Qm.n$ format, a number that is scaled by $2n$ bits. Similarly, it may also need $2m$ bits for the integer part. So, we cannot use a single precision integer of 32-bits to hold the result, we must use a double precision integer with 64-bits. This is the fixx_t data type. In line 5 we cast our inputs to 64-bits then multiply them. In line 8 we shift down N bits to get the answer to fit into a 32-bit integer. We'll skip line 7 for the moment.

Pictorially, the multiplication looks like Fig. 5.2 where we see the result of multiplying two 8-bit $Q3.4$ values using a 16-bit intermediate. The effect is the same when two 32-bit values are multiplied using a 64-bit intermediate. This figure shows clearly why the product must be shifted to the right to remove the extraneous lower-order bits. In the general case this shift is by N bits as in line 8 above. Finally, line 9 casts back to a 32-bit integer type, losing any overflow in the integer part of the fixed-point product, and returns.

We skipped line 7 above. This is the correction to remove truncation bias that was mentioned in Step 2 of the multiplication algorithm. We add a value which is equal to the highest bit position we will be discarding. This is much the same as adding

0.5 to a decimal number before keeping only the integer part so that the integer will round to the nearest value, not just the lowest value,

$$2068.34 \rightarrow \lfloor 2068.34 + 0.5 \rfloor = \lfloor 2068.85 \rfloor = 2068$$
$$2068.64 \rightarrow \lfloor 2068.65 + 0.5 \rfloor = \lfloor 2069.15 \rfloor = 2069$$

which is what we would expect from a round-to-the-nearest-integer operation.

Fig. 5.2 Multiplication of two fixed-point numbers. In this case two 8-bit fixed-point $Q3.4$ numbers are multiplied. The radix point is shown to separate the fractional bits from the integer bits (including the sign). The product requires 16-bits of which the middle eight are the desired answer as indicated. Fractional bits below this range are of lower order (smaller) and are discarded when the product is shifted down $n = 4$ bits. If any bits above the indicated range are set, and not the extended sign of the answer, then overflow has happened

What if we do check for overflow? Starting with addition and subtraction we have,

```
1   fix_t Q_add_over(fix_t x, fix_t y) {
2       uint32_t ux = x;
3       uint32_t uy = y;
4       uint32_t s = ux + uy;
5
6       if ((ux & MASK_SIGN) == (uy & MASK_SIGN)) {
7           if ((s & MASK_SIGN) != (ux & MASK_SIGN)) {
8               printf("Q_add_over: overflow!\n");
9               return 0;
10          }
11      }
12      return x+y;
13  }
14
```

```
15  fix_t Q_sub_over(fix_t x, fix_t y) {
16      uint32_t ux = x;
17      uint32_t uy = y;
18      uint32_t s = ux - uy;
19
20      if ((ux & MASK_SIGN) != (uy & MASK_SIGN)) {
21          if ((s & MASK_SIGN) != (ux & MASK_SIGN)) {
22              printf("Q_sub_over: overflow!\n");
23              return 0;
24          }
25      }
26      return x-y;
27  }
28
```

where in each case we return the same value as above, see lines 12 and 26, but
before that we use unsigned 32-bit integers to examine the sign of the result. We
use unsigned integers because overflow is undefined in the C standard for signed
integers. This tip we borrow from the *libfixmath* project.

For addition we know overflow happened if the signs of the arguments are the
same but the sign of the answer is not the same as the arguments. In line 6 we mask
the unsigned representation, same bit pattern as the signed representation, keeping
only the highest bit. This is the sign bit for the signed interpretation of the number. If
these are the same for both arguments, we then check whether the sign of the result
differs from the sign of the first argument (line 7). If they do, we know overflow
happened and we report it in line 8 and return zero in line 9. Naturally, a complete
implementation would set an error indicator and that would be used to know that the
return value is not valid.

The situation is similar for subtraction except instead of checking for the same
sign in the arguments line 20 checks to see if the signs are different. If so, we check
to see if the sign of the answer is the same as the sign of the first argument. If not,
we know overflow has happened and we report it in lines 22 and 23. Otherwise, we
return the difference in line 26 knowing that the two's-complement representation
will handle the signs properly.

Multiplication with overflow detection is,

```
1   fix_t Q_mul_over(fix_t x, fix_t y) {
2       fixx_t xx,yy,tt;
3       uint32_t ux = x;
4       uint32_t uy = y;
5
6       xx = (fixx_t)x;
7       yy = (fixx_t)y;
8       tt = xx * yy;
9       tt += 1<<(N-1);
10      tt >>= N;
11
12      if ((ux & MASK_SIGN) == (uy & MASK_SIGN)) {
13          if (((fix_t)tt & MASK_SIGN) != (ux & MASK_SIGN)) {
14              printf("Q_mul_over: overflow!\n");
15              return 0;
```

```
16|            }
17|        }
18|        return (fix_t)tt;
19| }
```

where we multiply as before but check our result starting at line 12. If the signs of
the arguments are the same (line 12) but the signs of the result and first argument
are different (line 13) we know overflow has happened and we report it in lines 14
and 15. If not, we return the product in line 18.

Lastly, we present division,

```
1| fix_t Q_div(fix_t x, fix_t y) {
2|     fixx_t xx,yy;
3|
4|     xx = (fixx_t)x << N;
5|     xx += (fixx_t)y / 2;
6|     return xx / (fixx_t)y;
7| }
```

which implements the algorithm for division outlined above. We cast the dividend
to a 64-bit integer and then scale it up by N bits in line 4. This sets xx to $x \times 2^n$
using 64-bits to hold the result. Line 5 is the truncation bias correction term which
adds a 64-bit version of y divided by 2. The actual division is performed in line 6
which as we saw above returns an already properly scaled quotient. Since the return
type for the function is fix_t the result of the division in line 6 is cast to the proper
32-bit size.

Before we finish this section let's take a quick look at the effect the truncation bias
correction term has on repeated multiplication. To do this we will look at powers of
$p = \pi - 3$ using $Q2.29$ which is capable of representing this value accurately to nine
decimal places. First, we calculate $(\pi - 3)^{10}$, then we calculate p^{10} with truncation
bias correction and without to show the effect of a series of ten multiplications in
a row,

$$(\pi - 3)^{10} = 0.00000000323897 \quad \text{floating-point value}$$
$$(\pi - 3)^{10} = 0.00000000372529 \quad Q2.29, \text{truncation bias correction}$$
$$(\pi - 3)^{10} = 0.00000000186265 \quad Q2.29, \text{no bias correction}$$

Notice how the truncation bias correction term keeps the final value reasonably close
to the actual value while without it the error accumulates and a serious loss of precision occurs. Clearly, truncation bias correction is an essential component of fixed-
point arithmetic.

5.3 Trigonometric and Other Functions

Now that we have implemented basic arithmetic operations in fixed-point, what
about trigonometric and other functions like log and exp? To implement these functions we have several options. We can use a look-up table, which is very fast, but

potentially either not accurate enough because there are not enough entries in the table, or it takes up too much memory because the table is big enough to be accurate. We can approximate the value by using a polynomial. Lastly, we can approximate the value by summing a Taylor series expansion. Let's look at each of these options for the sine function. We can calculate the cosine from the sine and once we have both sine and cosine we can calculate the tangent. We will ignore the inverse trigonometric functions. After this, we will show implementations for square root, logarithm and exponential functions.

Trigonometric functions. For our fixed-point library we will implement the three basic trigonometric functions: sine, cosine, and tangent. All of these will be based on the sine which we will implement in three different ways so that we can compare the performance and accuracy of each method. The cosine is, of course, easy to derive from the sine since,

$$\cos(\theta) = \sin(\frac{\pi}{2} - \theta)$$

and, by definition,

$$\tan \theta = \frac{\sin \theta}{\cos \theta}$$

meaning we get everything once we have the sine implemented. We will implement the sine using a table, a Taylor series expansion, and a polynomial fit. First the table,

```
 1 fix_t Q_sin_table(fix_t x) {
 2     int32_t sign;
 3     fix_t angle;
 4
 5     if (x < 0) x += PId;
 6     sin_val_sign(&x, &sign);
 7
 8     angle = Q_mul(x,r2d);
 9     angle += 1 << (N-1);
10     return sign*sin_tbl[angle >> N];
11 }
```

which uses the globally defined `sin_tbl` array defined above with `Q_init`. This function also uses a helper function, `sin_val_sign`, which we describe below. This function will be used by our other sine implementations, too. It takes an angle, in radians, and updates it to be in the range $[0, \pi/2]$ along with a sign. This is equivalent to paying attention to which quadrant the angle puts us in and then using the appropriate angle and sign for that quadrant to give us the right sine value. Recall that `Q_init` initializes `sin_tbl` with the sine for each degree from $[0, 90]$. This means that we only need consider input angles in that range.

For `Q_sin_table` we first see if the argument is negative. If so, we add 2π to make it positive. So, the input to the function must be in the range $[-2\pi, 2\pi]$. This is line 5. Note that we use normal addition thereby making an implicit assumption that we will not overflow (nor to we check for it). In line 6 we call `sin_val_sign` to adjust our input angle to be in the range $[0, \pi/2]$ and set `sign` accordingly. In line 8 we convert our adjusted input angle, assumed to be in radians, to degrees.

This is because we built the table by degrees and want to index it by degree. Line 9 adds 0.5 to the angle, this is the bias correction, as we are interested in mapping angle to the nearest degree so we can use the look-up table. In line 10 we truncate angle, dropping the fractional part by shifting it down N bits. This means that angle >> N is an integer in the range $[0, 90]$ degrees. With this integer we index our sine table and, after multiplying by sign, return the final answer. This function is fast but has the overhead of building the table in the first place for each call to Q_init and the memory used by the table. Additionally, while most accurate for inputs which have no fractional component, it is not as accurate as it could be for arbitrary inputs. The exercises will challenge the reader to increase the accuracy by linear interpolation.

Let's look at sin_val_sign which we used above and will use again,

```
 1  void sin_val_sign(int32_t *angle, int32_t *sign) {
 2
 3      if ((*angle >=0) && (*angle <= f90)) {
 4          // quadrant I
 5          *sign = 1;
 6      } else {
 7          if ((*angle > f90) && (*angle <= f180)) {
 8              // quadrant II
 9              *sign = 1;
10              *angle = f180-*angle;
11          } else {
12              if ((*angle > f180) && (*angle <= f270)) {
13                  // quadrant III
14                  *sign = -1;
15                  *angle = *angle-f180;
16              } else {
17                  // quadrant IV
18                  *sign = -1;
19                  *angle = f360-*angle;
20              }
21          }
22      }
23  }
```

This function accepts as input via pointers the current angle and a sign value which it will set. By comparing with the constants f90, f180, f270, and f360, each of which is set in Q_init, the function determines which quadrant angle falls in. It then sets sign appropriately for that quadrant and updates the value of angle so that the sine of the new angle, times sign, will be the proper sine for the original angle. This means we only ever work with angles in the range $[0, \pi/2]$.

We get Q_cos_table and Q_tan_table for free via Q_sin_table,

```
1  fix_t Q_cos_table(fix_t x) {
2      return Q_sin_table(Q_sub(PIh,x));
3  }
4
5  fix_t Q_tan_table(fix_t x) {
6      return Q_div(Q_sin_table(x), Q_cos_table(x));
7  }
```

where we have made use of the trigonometric identities given above. Note the use of the constant PIh for $\pi/2$ which is set in Q_init for the given $Qm.n$ fixed-point representation.

Now we consider implementing sine using a Taylor series expansion. Please see any calculus book for a description of a Taylor series. Here we simply use the well-known series expansion for $\sin x$,

$$\sin x = x - \frac{x^3}{3!} + \frac{x^5}{5!} - \frac{x^7}{7!} + \frac{x^9}{9!} + \cdots$$

after mapping the input angle to $[0, \pi/2]$ as before. This leads to,

```
 1  fix_t Q_sin_taylor(fix_t x) {
 2      fix_t x3,x5,x7,x9;
 3      fix_t ans;
 4      int32_t sign;
 5
 6      if (x < 0) x += PId;
 7      sin_val_sign(&x, &sign);
 8
 9      x3 = Q_mul(Q_mul(x,x),x);
10      x5 = Q_mul(Q_mul(x3,x),x);
11      x7 = Q_mul(Q_mul(x5,x),x);
12      x9 = Q_mul(Q_mul(x7,x),x);
13
14      ans = x;
15      if (x3 != 0) ans -= Q_mul(st1,x3);
16      if (x5 != 0) ans += Q_mul(st2,x5);
17      if (x7 != 0) ans += Q_mul(st3,x7);
18      if (x9 != 0) ans += Q_mul(st4,x9);
19
20      return sign*ans;
21  }
```

where lines 6 and 7 check for a negative input, adding 2π if it is, and we map the argument to $[0, \pi/2]$ while setting sign. Lines 9 through 12 calculate powers of the input, x. We are implementing the series directly to make it as easy as possible to see the translation from the equation to the code so we calculate all the powers we need in advance. A little algebraic manipulation will show that we only really need x and x^2 plus some well-placed parentheses. See the exercises for another challenge to the reader. Line 14 sets ans to the first term in the series. Lines 15 through 18 add subsequent terms if the power of x for that term is not zero. This may happen if $x < 1$ and we do not have enough precision in the n fractional bits of the $Qm.n$ number to represent the power. The magic constants st1, st2, st3, and st4 are the reciprocals of the factorials in the series representation. These are calculated in advance for the given Q representation when Q_init is called. Finally, line 20 multiplies by the sign and returns the final estimate for $\sin x$. As before, we get $\cos x$ and $\tan x$ for free,

```
1  fix_t Q_cos_taylor(fix_t x) {
2      return Q_sin_taylor(Q_sub(PIh,x));
3  }
4
5  fix_t Q_tan_taylor(fix_t x) {
6      return Q_div(Q_sin_taylor(x), Q_cos_taylor(x));
7  }
```

where the form of the functions matches the table versions above but is based on Q_sin_taylor instead.

Our final $\sin x$ implementation comes from a set of points calculated from a floating-point version of sine that is then curve fit to a 5-th degree polynomial. We calculated $\sin x$ for 100 equally spaced points from $[0, \pi/2]$ using the C library version of $\sin$. We then fit a 5-th degree polynomial to these points to find the coefficients p_0, p_1, p_2, p_3, p_4, and p_5 which minimize the squared difference between $\sin x$ and $f(x)$ where,

$$f(x) = p_0 + p_1 x + p_2 x^2 + p_3 x^3 + p_4 x^4 + p_5 x^5$$

for each of the 100 data points x_i, $i = 0, 1, \ldots, 99$. This is standard least-squares curve fitting. We used the Python *numpy* package [3] to calculate the coefficients (see *polyfit*). These coefficients are mapped to the current Q notation in Q_init and set to the variables p0, p1, p2, p3, p4, and p5, respectively. With these parameters we are able to approximate the sine in the range $[0, \pi/2]$ with,

```
1  fix_t Q_sin_poly(fix_t x) {
2      fix_t x2,x3,x4,x5;
3      fix_t ans;
4      int32_t sign;
5
6      if (x < 0) x += PId;
7      sin_val_sign(&x, &sign);
8
9      x2 = Q_mul(x,x);
10     x3 = Q_mul(x2,x);
11     x4 = Q_mul(x3,x);
12     x5 = Q_mul(x4,x);
13
14     ans = p0 + Q_mul(p1,x) + Q_mul(p2,x2) + Q_mul(p3,x3) +
15           Q_mul(p4,x4) + Q_mul(p5,x5);
16     return sign*ans;
17 }
```

where as before, lines 6 and 7 adjust for negative arguments and map x to $[0, \pi/2]$ while setting sign. Lines 9 through 12 calculate the powers of x for the 5-th degree polynomial. Lines 14 and 15 directly implement the polynomial using Q_mul to multiply with a proper intermediate number of bits and ordinary addition thereby ignoring any potential overflow. Lastly, line 16 multiplies by sign and returns the answer. As before, we can write $\cos x$ and $\tan x$ using the polynomial sine,

```
1 fix_t Q_cos_poly(fix_t x) {
2     return Q_sin_poly(Q_sub(PIh,x));
3 }
4
5 fix_t Q_tan_poly(fix_t x) {
6     return Q_div(Q_sin_poly(x), Q_cos_poly(x));
7 }
```

Comparison of trigonometric functions. All three sets of trigonometric functions implemented above have a version of sine at their heart. Therefore, we can compare them by looking at the accuracy of the sine functions. If we generate the sine, using the table, series expansion, or polynomial fit, for 100 randomly selected angles in $[0, 2\pi)$ we can compare the resulting values to the actual floating-point sine. Any deviation can be used to assess the quality of the sine approximation.

If we generate the table with code like this,

```
 1 Q_init(7);
 2 srand(time(NULL));
 3
 4 for(i=0; i<100; i++) {
 5     f = 2*3.14159265*((double)rand()/RAND_MAX);
 6     a = Q_to_fixed(f);
 7     printf("%0.8f %0.8f %0.8f %0.8f %0.8f\n", f, sin(f),
 8         Q_to_double(Q_sin_table(a)),
 8         Q_to_double(Q_sin_taylor(a)),
 9         Q_to_double(Q_sin_poly(a)));
10 }
```

where we include the stdlib.h C library and use srand to seed the random number generator which we call in line 5 to set f to a random value in the range $[0, 2\pi)$. We note that it is well known that rand is a very poor pseudo-random number generator and that we have a very (very) small chance of selecting 2π as rand returns a number from $[0, RAND_MAX]$ and we are dividing by RAND_MAX. However, for our purposes we accept the inefficiency and take the risk. The reader is encouraged, however, to develop an understanding of pseudo-random number generation, which is fascinating, by referring to one of the many good texts on the subject. For example, see Gentle's *Random Number Generation and Monte Carlo Methods* [4].

Now that we have our randomly selected sine values we can calculate the delta between the floating-point value and the three approximations for each angle and then report the absolute maximum deviation ($|\Delta|$), the mean ($\bar{x}$), and the standard deviation (σ) of the deltas. These will give us a feel for how the functions perform over the their full range. The results of this analysis for $Q7.24$ numbers are in Table 5.1.

We see that while all three approximations are adequate with relatively small errors, on average, it is clear that Q_sin_table is the least accurate, as expected because it forces the argument to the nearest degree, and that Q_sin_poly is to

Function	$|\Delta|$	$\bar{x}$	σ
Q_sin_table	8.4×10^{-3}	-6.5×10^{-5}	3.8×10^{-3}
Q_sin_taylor	8.2×10^{-3}	-7.8×10^{-6}	8.2×10^{-3}
Q_sin_poly	7.7×10^{-6}	-1.1×10^{-7}	3.9×10^{-6}

Table 5.1 Comparison of the difference between the floating-point sine and each of the fixed-point sine functions using summary statistics for 100 randomly selected angles over the entire range $[0, 2\pi)$

be preferred over Q_sin_taylor in terms of accuracy. If we count the number of multiplications we see that Q_sin_poly is also more efficient than Q_sin_taylor with nine multiplications to as many as twelve.

Square root and transcendental functions. In addition to basic trigonometric functions we want to add square root, natural logarithm and exponential functions to our library. Let's look at how we can implement these functions starting first with square root.

Perhaps the fastest way to implement a fixed-point square root function is to use Newton's method. This is an iterative method to find the roots of a function by solving $f(x) = 0$. The method selects an initial guess for the root, x_0, and then iterates,

$$x_{i+1} = x_i - \frac{f(x_i)}{f'(x_i)}$$

until convergence where $f'(x)$ is the first derivative of $f(x)$. In this case, we want $f(x) = x^2 - n = 0$ which will find the value of x which when squared equals a specific number n. This means that x will be equal to the square root of n which is exactly what we want. If we work through the steps we have,

$$f(x) = x^2 - n$$
$$f'(x) = 2x$$
$$x_{i+1} = x_i - \frac{f(x_i)}{f'(x_i)}$$
$$= x_i - \frac{x_i^2 - n}{2x_i}$$
$$= x_i - \frac{1}{2}\left(x_i - \frac{n}{x_i}\right)$$
$$= \frac{1}{2}\left(x_i + \frac{n}{x_i}\right)$$

which implies that we need only an initial guess followed by some number of iterations of the final equation above to generate a good approximation to the square root of n. It is known that this method converges quickly so we only need a handful of iterations.

Translating this into code gives,

```
 1  fix_t Q_sqrt(fix_t n) {
 2      fix_t a;
 3      int32_t k;
 4
 5      if (n <= 0) return -(1<<N);
 6      if (n == (1<<N)) return (1<<N);
 7
 8      a = n >> 1;
 9
10      for(k=0; k < 10; k++)
11          a = Q_mul(1 << (N-1),Q_add(Q_div(n,a),a));
12
13      return a;
14  }
```

where we first check for negative arguments and return -1 if we have one in line 5. Recall that our fixed-point numbers are binary fractions scaled by N bits so shifting 1 up N bits gives us a value equal to one in our current Q notation. Line 6 is a quick check to see if our argument is one. If so we immediately return the square root (1). Line 8 is our initial guess which we store in a. We are looking for a square root so our initial guess is $n/2$ which is exactly what we get by shifting the argument one bit position to the right. The loop in lines 10 and 11 performs the actual iteration. We fix the number of iterations at ten though a more sophisticated approach would perhaps be sensitive to the number of bits in the fractional part of our numbers and adjust the iteration limit accordingly. Line 11 is a direct translation of our final equation above using our previously defined arithmetic functions. Note that for simplicity we are not checking for any overflow conditions. Also note the use of 1 << (N-1) which is exactly 0.5 for N bits in the fractional part. Finally, line 13 returns a which is the approximate square root of n.

Repeating a random analysis similar to the one performed above for the sine functions we can compare Q_sqrt to the floating-point sqrt function for 100 randomly selected values from $[1, 1000)$ using $Q10.21$ format numbers. If we do this we get the following results,

$|\Delta|$	$\bar{x}$	σ
3.6×10^{-7}	-1.3×10^{-7}	1.4×10^{-7}

which indicates that this approach gives quite accurate answers in general.

We consider the exponential function, e^x, next as we will need this function in order to efficiently implement the natural logarithm. The Taylor series expansion of e^x is particularly well suited for implementation on a computer. The series itself is,

$$e^x = 1 + x + \frac{x^2}{2!} + \frac{x^3}{3!} + \frac{x^4}{4!} + \cdots = \sum_{i=0}^{\infty} \frac{x^i}{i!}$$

Looking at the series shows that if the current term in the series is t_i then the next term in the series is $t_{i+1} = t_i(\frac{x}{i})$ which means that we need not calculate the factorials completely but will get them by multiplying the current term by $\frac{x}{i}$ and adding it to a running total. If we accumulate terms until we get a term that is below a given threshold, say too small to add anything meaningful to the total given the number of bits used for the fractional part of the fixed-point number, we have a particularly nice implementation. Writing this in C using EXP_TOLERANCE to be the cut-off value for our terms, 0.000001 or the like, and adding a maximum number of iterations (EXP_ITER_MAX = 50) gives the following,

```
 1  fix_t Q_exp(fix_t x) {
 2      int32_t i;
 3      fix_t e, c, t, tol;
 4
 5      tol = Q_to_fixed(EXP_TOLERANCE);
 6      i = 0;
 7
 8      c = 1 << N;
 9      e = 1 << N;
10      t = x;
11
12      do {
13          e = Q_add(e,t);
14          c = Q_add(c, 1<<N);
15          t = Q_mul(t, Q_div(x,c));
16          i++;
17      } while ((abs(t) > tol) && (i < EXP_ITER_MAX));
18
19      return e;
20  }
```

where line 5 sets the tolerance and line 6 initializes the iteration counter. Line 8 sets c to one. This is the term counter and is set to one because the first term in the series (1) is used to initialize the accumulator, e in line 9. The current term is then the second term in the series, which is $x = x^1/1!$, and we initialize t with it. The loop in lines 12 through 17 adds new terms to the accumulator (line 13) while bumping the term counter (line 14). Line 15 calculates the next term in the series from the previous term and sets t accordingly. Line 16 increments the iteration counter while line 17 checks to see if we can exit the loop by either calculating a term below the tolerance or by reaching the limit on iterations. At this point e contains the sum of the series, which is e^x, and it is returned in line 19.

Repeating the random value test for 100 random arguments in the range $[-4, 4)$ for $Q10.21$ format numbers gives,

$$\begin{array}{ccc} |\Delta| & \bar{x} & \sigma \\ \hline 2.0 \times 10^{-5} & -9.2 \times 10^{-7} & 4.5 \times 10^{-6} \end{array}$$

showing that Q_exp does give good results.

With e^x implemented we can now add our final library function, the natural log ($\log x$) which is implemented using Newton's method just as square root was. In this case, our function is $f(x) = e^x - c = 0$ which is zero when x is the log of c. The first derivative of this function is $f'(x) = e^x$ which means our iteration function is,

$$x_{i+1} = x_i - \frac{f(x_i)}{f'(x_i)} = x_i - \frac{e^{x_i} - c}{e^{x_i}} = x_i - 1 + ce^{-x_i}$$

where we again set our initial guess to $x_0 = c/2$. Translating to C code gives,

```
 1  fix_t Q_log(fix_t x) {
 2      fix_t a;
 3      int32_t k;
 4
 5      if (x <= 0) return -(1<<N);
 6      if (x == (1<<N)) return 0;
 7
 8      a = x >> 1;
 9
10      for(k=0; k < 7; k++) {
11          a = Q_add(a, Q_mul(x,Q_exp(-a)));
12          a = Q_sub(a,1<<N);
13      }
14
15      return a;
16  }
```

where a sanity check is made in line 5 to disallow negative arguments (return -1) and a quick check for a known value ($\log 1 = 0$) is in line 6. Line 8 sets a to our initial guess of $c/2$. Lines 10 through 13 are the iteration loop. This is why we needed to implement e^x before we could implement the logarithm. Line 11 calls Q_exp as part of the update to a. Line 12 subtracts one from the sum in line 11. The number of iterations is hard-coded to seven which gives good results. Again, a more sophisticated approach would be aware of the number of bits in the fractional part of our fixed-point numbers and set the loop limit accordingly. Line 15 returns a as our approximation of the natural log.

A final repeat of the random value test with 100 random arguments from $[0.5, 10)$ gives,

$|\Delta|$	$\bar{x}$	σ
1.3×10^{-5}	4.0×10^{-7}	4.4×10^{-6}

again showing good agreement with the floating-point C library function.

At this point our C library of fixed-point functions is complete. We now turn our attention from the *what* of fixed-point numbers to briefly consider the questions of *why* and *when*.

5.4 When to Use Fixed-Point Numbers

Perhaps the most obvious time to use fixed-point numbers is when working on a system that does not have floating-point capability in hardware. This includes many small systems, like microcontrollers and digital signal processors (DSP), but also includes older computers that lack floating-point hardware. Using fixed-point numbers on these systems makes sense and is much faster than using software to emulate floating-point hardware. For example, older Macintosh computers based on the Motorola 68000 and later processors often used software to emulate floating-point hardware. Called SANE (Standard Apple Numerics Environment), it implemented IEEE 754 floating point arithmetic in software for early Macintosh computers that lacked the floating-point hardware found in the Motorola 68040 or did not use the 68881 floating-point coprocessor. The SANE library provided maximum numeric compatibility ensuring that floating-point operations on one computer matched those of another that used hardware instead of software for the calculations. However, the simplicity of fixed-point arithmetic would have greatly improved performance on these early systems.

In 1993, Id Software released DOOM [5]. This game was one of the earliest first-person shooter (FPS) video games. FPS games give the user the illusion of being in a three-dimensional environment and as such require a lot of geometric graphics processing at high speed to enable realistic (limited for DOOM, granted) rendering of the environment. However, in 1993 only Intel 80486DX microprocessors had built-in hardware floating point abilities. As processor speeds were very low compared to current standards, only tens of megahertz, it was crucial that calculations be as fast as possible while still using only software. Id's solution was to use fixed-point arithmetic for all graphics calculations. Specifically, they used $Q15.16$ format numbers which fit in a 32-bit signed integer.

Another reason to consider fixed-point math is power consumption. The floating-point unit on processors uses more power than the integer unit hence fixed-point arithmetic will use less power than the same calculations performed with hardware floating-point. This is particularly important in mobile applications.

Lastly, many popular audio and video codecs are designed to be implemented using fixed-point arithmetic so that simple devices that lack floating-point hardware can be used. For example, the open source Shine project [6] is a fast fixed-point MP3 encoder which can be used on simple computers like the Raspberry Pi.

5.5 Chapter Summary

In this chapter we introduced Q notation to specify fixed-point numbers. These numbers interpret an integer as a floating point number with a fixed number of fractional bits. We learned how to convert floating-point numbers to fixed-point and back. We also implemented arithmetic operations in C for fixed-point numbers using signed 32-bit integers. Additionally, we implemented trigonometric functions in three different ways and compared them for accuracy. To round out the library we then added square root, natural logarithm, and exponential functions. Lastly, we talked about when to use fixed-point numbers and gave some examples of places where fixed-point numbers have been used.

Exercises

5.1. The function Q_sin_table rounds its argument to the nearest degree and then returns the sine stored in the precomputed table for that degree. This can be made more accurate by linearly interpolating between the degrees that bound the input angle. For example, if the input value, as a fixed-point number in some $Qm.n$ representation, is equivalent to 45.4 degrees then Q_sin_table will round down to 45 degrees and return 0.70710678 as the sine. The actual sine of 45.4 degrees is 0.71202605 so the returned value underestimates the true value.

In linear interpolation the values that bound the input, in this case $45 < 45.4 < 46$ degrees, are used to determine the line between the values, $y = mx + b$, with $m = (y_2 - y_1)/(x_2 - x_1)$ and $b = y_1 - mx_1$ where $x_1 = 45$, $x_2 = 46$, $y_1 = \sin(x_1)$ and $y_2 = \sin(x_2)$. Then, $\sin(45.4)$ is equal to $m(45.4) + b$.

In this case, then, the improved answer is $\sin(45.4) = m(45.4) + b = 0.71199999$ with $m = 0.01223302$ and $b = 0.15662088$ for a difference on the order of 2.6×10^{-5} when compared to the true result. This is better than the difference of 4.9×10^{-3} by simply rounding to the nearest degree. Add linear interpolation to Q_sin_table. **

5.2. Add a function, Q_ipow, to the library that takes two arguments. The first is a fixed-point number, x, and the second is an integer, n, which may be negative. The return value of the function is x^n. *

5.3. Add a function, Q_pow, to the library that takes two arguments. The first is a fixed-point number, x, and the second is also a fixed-point number, y. The return value of the function is x^y. (Hint: $x^y = e^{y \log(x)}$) *

5.4. Add a function, Q_asin, to the library that takes a sine value, positive and negative, and returns the angle for that sign, in radians. Use a Taylor series expansion,

$$\sin^{-1} x = x + \left(\frac{1}{2}\right)\left(\frac{x^3}{3}\right) + \left(\frac{1}{2}\right)\left(\frac{3}{4}\right)\left(\frac{x^5}{5}\right) + \left(\frac{1}{2}\right)\left(\frac{3}{4}\right)\left(\frac{5}{6}\right)\left(\frac{x^7}{7}\right) + \cdots$$

where you may need terms to x^9 or higher. (Hint: the coefficient of term t_{i+1} can be created from t_i) **

5.5. Create a Python class, Fixed, that duplicates the arithmetic functionality of the C fixed-point library. Be sure to overload the operators for $+$, $-$, $/$, $*$ so that the fixed-point class can be used in expressions. **

5.6. The function Q_sin_taylor uses terms out to x^9 and calculates them explicitly for input x. Show via algebraic manipulation that the same truncated series sum can be found using only x and x^2. Then, update Q_sin_taylor to use this form. Note that you will need to introduce new constants in place of the values used for $st1$ through $st4$. Update Q_init as well to set these new values. **

References

1. TMS320C64x DSP Library Programmer's Reference Literature Number: SPRU565B, Texas Instruments, October 2003.
2. https://code.google.com/p/libfixmath/.
3. http://www.numpy.org/.
4. Gentle, J. E. Random number generation and Monte Carlo methods. Springer, 2003.
5. Doom. Id Software. 1993. Video game.
6. https://github.com/savonet/shine.

Chapter 6
Decimal Floating Point

Abstract Decimal floating-point is an emerging standard which uses base 10 instead of base 2 to represent floating-point numbers. In this chapter we will take a look at how decimal floating-point numbers are stored using the IEEE 754-2008 standard as our reference. We will then look at a C language software implementation of decimal floating-point which conforms to the IEEE standard and give some examples of its use. We follow this with a look at the Python `decimal` library. We end with some thoughts on decimal floating-point and its use.

6.1 What Is Decimal Floating-Point?

Modern computers make extensive use of floating-point arithmetic. Virtually all of this is accomplished with binary floating-point numbers. Binary floating-point numbers use base 2 and generally represent their numbers as a significand (mantissa), also binary, multiplied by 2 raised to some exponent. The standard for this type of floating-point number is given in the IEEE 754-2008 document [1] which has been universally adopted and includes hardware implementations for maximum performance. However, it is known that there are often accuracy issues with binary floating-point because of rounding errors or lack of precision when converting decimal numbers to binary. For example, decimal 0.1 is an infinitely repeating binary fraction which therefore must be truncated when stored in a finite number of bits.

Decimal floating-point (DFP) was introduced to address these issues. In DFP, the base is 10, not 2, so that decimal numbers may be represented exactly (to within the number of digits in the significand). The IEEE standard that defines binary floating-point also defines decimal floating-point formats and operations. This is the standard we will refer to in this chapter. Even though there is limited hardware support for DFP, notably from IBM, we will focus exclusively on software implementations here.

© Springer International Publishing Switzerland 2015   181
R.T. Kneusel, *Numbers and Computers*, DOI 10.1007/978-3-319-17260-6_6

A binary floating-point number is represented as,

$$\pm d_0.d_1 d_2 d_3 \ldots d_{p-1} \times 2^e, \quad 0 \le d_i < 2$$

where $d_0.d_1 d_2 d_3 \ldots d_{p-1}$ is the p digit *significand* (or *mantissa*), 2 is the base and e is the integer exponent. Similarly, a decimal floating-point number is represented as,

$$\pm d_0.d_1 d_2 d_3 \ldots d_{p-1} \times 10^e, \quad 0 \le d_i < 10$$

where the digits of the significand are decimal digits. This implies, as we will be storing our DFP numbers on a computer, that there is some sort of encoding for the significand so the digits $0 \ldots 9$ can be stored efficiently. This is not necessary in a binary floating-point number. For a description of the difference between a floating-point number, regardless of base, and a real number see the introduction to Chap. 3.

6.2 The IEEE 754-2008 Decimal Floating-Point Format

In this section we introduce decimal floating-point using the IEEE 754 format. We will describe the way in which DFP numbers are stored, using the fixed *decimal32*, *decimal64* and *decimal128* storage formats, and cover special values such as zero, infinity, and Not-a-Number (NaN). These values have direct analogues in binary floating-point.

Storage Formats. The IEEE 754 standard specifies DFP numbers as a sign, coefficient and exponent so that the actual number is,

$$n = (-1)^{sign} \times coefficient \times 10^{exponent}$$

with the sign a single bit, 0 if positive and 1 for negative, and the exponent in excess binary notation. Excess notation means that the actual exponent is found by subtracting the bias from the stored value. There are two equivalent storage formats for the coefficient field, one as a binary integer and the other as a packed decimal format. In this chapter we will be working with a software library which uses the packed decimal format so that is the format we will concentrate on. The few existing hardware implementations of DFP use the packed format most often as well.

While the actual DFP value is specified by the sign, coefficient and exponent the actual way in which DFP numbers are stored in memory is fairly complicated. The *decimal64* format is made up of four different fields,

S	CF	BXCF	CCF
1	5	8	50

where S is the sign (1 bit), CF is the combination field (5 bits), $BXCF$ is the biased exponent continuation field (8 bits), and CCF is the coefficient continuation field

(50 bits). For a *decimal32* number the *BXCF* is 6 bits and the *CCF* is 20 bits. For *decimal128* these fields are 12 bits and 110 bits respectively.

Let's start by decoding an example of a *decimal64* number, in this case 0.1. While 0.1 is infinitely repeating when expressed in binary in decimal it is simply a single digit. If we encode 0.1 as a DFP number it will use eight bytes of memory (64 bits). We can then dump each of these bytes as a sequence of hexadecimal numbers,

$$0.1 \to 22 \ 34 \ 00 \ 00 \ 00 \ 00 \ 00 \ 01$$

from which we see that the fields are, in binary,

```
S    = 0
CF   = 01000
BXCF = 10001101
CCF  = 0000000000 0000000000 0000000000 0000000000 0000000001
```

where we have written the *CCF* as five groups of ten binary digits. We'll see why momentarily. For now, let's consider the other fields. The sign is straightforward enough. It is zero so the number is positive since $(-1)^0 = 1$. The *CF* field requires some explanation. It is a five bit field which encodes the two leftmost binary digits of the bias exponent along with the leftmost digit of the coefficient. There are ten possible coefficient digits and three different values of the leftmost two bits of the exponent for a total of 30 possible five bit combinations. As $2^5 = 32$ this means that 30 of the possible bit patterns encode the values we want and we still have two left over for other uses as we will see below. Table 6.1 shows all the bit patterns for the *CF* field and how to interpret them to get the exponent and coefficient digit values.

Digit	00	01	10
0	00000	01000	10000
1	00001	01001	10001
2	00010	01010	10010
3	00011	01011	10011
4	00100	01100	10100
5	00101	01101	10101
6	00110	01110	10110
7	00111	01111	10111
8	11000	11010	11100
9	11001	11011	11101

Table 6.1 Decoding the *CF* field. Locate the *CF* bit pattern to read off the leftmost two bits of the biased exponent (column) and the leftmost digit of the coefficient (row). After [2]

For our example we have 01000 which corresponds to exponent bits 01 and a leftmost coefficient digit of 0.

The remainder of the ten bit exponent is in the *BXCF* field. Combining this with the decoded *CF* value gives $0110001101_2 = 18D_{16} = 397$ for the biased exponent. The bias value for *decimal64* is 398 so the unbiased exponent is $397 - 398 = -1$.

The bias value for *decimal32* is 101 while the bias for *decimal128* is 6176. The minimum and maximum exponents for each of the three encodings is therefore,

	Minimum	Maximum
decimal32	−101	90
decimal64	−398	398
decimal128	−6176	6111

The last part we need to decode is the coefficient field itself. We already know the leftmost digit of the coefficient field from the *CF* field. It is zero. The remainder of the coefficient is encoded in the *CCF* field. This field encodes the decimal digits of the coefficient in a packed format called *densely packed decimal* or DPD. This format uses ten bits to encode three decimal digits. This is more efficient than classical binary-coded decimal which would require 12 bits to encode three decimal digits See Chap. 2 for an explanation of binary-coded decimal.

The DPD encoding takes ten bits labeled *pqr stu v wxy* and maps them to three decimal digits represented in binary as *abcd efgh ijkm* where each letter is a binary digit (*l* is skipped to avoid confusion with 1). This group of ten bits is called a *declet* and can be decoded by mapping through Table 6.2.

vxwst	abcd	efgh	ijkm
0----	0pqr	0stu	0wxy
100--	0pqr	0stu	100y
101--	0pqr	100u	0sty
110--	100r	0stu	0pqy
11100	100r	0pqu	100y
11110	0pqr	100u	100y
11111	100r	100u	100y

Table 6.2 Unpacking a declet (*pqr stu v wxy*) by translating it into three groups of binary digits representing three decimal digits (*abcd efgh ijkm*). After [2]

For our example we have five declets which corresponds to 15 decimal digits, three digits per declet, plus one additional leading (leftmost) digit of zero. This means that a *decimal64* number has 16 digits of precision. For *decimal32* the precision is 7 digits while for *decimal128* it is 34 digits. Four of the five declets in our example are all zeros mapping to three zero digits. The rightmost declet is 0000000001 which means that in Table 6.2 *v* is zero corresponding to the first row of the table. Therefore, the three decimal digits of this last declet are,

$$0pqr\ 0stu\ 0wxy$$
$$0000\ 0000\ 0001$$

corresponding to 001 in decimal. So, finally, we have decoded the entire DPF number: sign, coefficient, and exponent. The final value is, as we expect,

$$n = (-1)^0 \times 000000000000001 \times 10^{-1} = 1 \times 10^{-1} = 0.1$$

It should be noted that, unlike IEEE binary floating point, DFP does not have a single exponent for each number. This means that there are possibly multiple representations of the same number since $1 \times 10^{-1} = 10 \times 10^{-2}$, etc. The standard defines a preferred exponent, based on the arguments to an operation, in order to produce consistent output. The set of representations mapping to the same number are called its *cohort*.

Packing numbers into a DPD declet is straightforward. For the example above we produced three BCD digits (four bits each, the value in binary is the value of the digit) and called the bits of each *abcd*, *efgh*, and *ijkm*. Using these labels for the bits we can take any set of three BCD digits and encode them into a DPD declet using Table 6.3. Recall that the ten bits of the declet are labeled *pqr stu v wxy* which can be read from the table by finding the row matching the *aei* bit values from the input BCD numbers. We will not work an example here but save using Table 6.3 for the exercises.

aei	*pqr*	*stu*	*v*	*wxy*
000	bcd	fgh	0	jkm
001	bcd	fgh	1	00m
010	bcd	jkh	1	01m
011	bcd	10h	1	11m
100	jkd	fgh	1	10m
101	fgd	01h	1	11m
110	jkd	00h	1	11m
111	00d	11h	1	11m

Table 6.3 Packing three BCD numbers (*abcd efgh ijkm*) into a declet (*pqr stu v wxy*). Locate the row by matching the first bit of each BCD number to the value under the *aei* column, then, the bits of the declet are read from the remaining columns of that row. After [2]

Special Values. Just as IEEE 754 defines special values for binary floating-point, so it does for decimal floating-point. These are zero, which may be signed or unsigned but interpreted as zero regardless, infinities, and Not-a-Number (NaN). Let's consider each of these.

For *decimal64* the eight bytes representing +0 are 22 38 00 00 00 00 00 00. Parsing this gives S of 0, CF of 01000, and $BXCF$ of 10001110. As before, the CF implies that the first two bits of the exponent field are 01 while the first digit of the coefficient field is zero. Glancing at the remaining digits of the bytes representing zero makes it clear that the entire coefficient field is zero. What about the exponent? Combining the 01 from the CF field with the bits of the $BXCF$ field gives,

$$e = 0110001110_2 = 18E_{16} = 398$$

which unbiased is $e = 398 - 398 = 0$. So, zero is represented in DFP as $(-1)^0 \times 0 \times 10^0$ which is as we would expect. What happens if we consider -0? The only value that changes is the sign giving $(-1)^1 \times 0 \times 10^0$, which is still zero.

If a number is too large, positive or negative, to be represented, a special value of infinity is used. To signal infinity the CF field is set to one of the two 5-bit values that are not used to encode the exponent. In this case, the *CF* field will be 11110. The sign bit determines whether the value is $+\infty$ or $-\infty$. Other bits are ignored. So, if we try to represent 1×10^{1000} with a *decimal64* number, which is too large since the largest allowed exponent value is 369, we instead get infinity,

$$78 \ \ 00 \ \ 00 \ \ 00 \ \ 00 \ \ 00 \ \ 00 \ \ 00$$

with,

$$S \ = 0$$
$$CF = 11110$$

and *BXCF* and *CCF* all zero. Negative infinity is the same with *S* set to 1.

The IEEE 754 standard for binary floating-point defines Not-a-Number values to represent attempts to encode values that are not legitimate numbers. NaNs are also defined for decimal floating-point by the special *CF* field value of 11111. For example, attempting to calculate the logarithm of a negative number will result in a NaN. In addition to the bits of the *CF* field, the leftmost bit of the *BXCF* field is used to determine whether or not the NaN is signaling (bit set to 1) or quiet (bit set to 0). A signaling NaN will throw a floating-point exception while a quiet NaN will not.

When a NaN is generated the IEEE 754 standard allows implementations to use the *CCF* field to encode information about the source of the NaN. This is called its payload. Furthermore, the standard also ensures that subsequent operations involving a NaN will propagate the payload to the result of the operation. It appears, in practice, that this ability is seldom used. Instead, the appearance of the NaN is taken to mean there is an error in the expression or code without using the payload to provide any additional information. See Chap. 3 for an example using the payload for a binary floating-point number.

Rounding Modes. The IEEE standard defines multiple rounding modes. The standard assumes that operations are performed with "infinite" precision and are then fit into the storage format at hand. This will require rounding of results so that the proper number of output digits can be set. These are the rounding modes. Typically, users seldom modify these modes but there are times when it is beneficial to do so (interval arithmetic is one such instance). There are five rounding modes that can apply to decimal floating-point numbers. These are,

Name	Description
roundTowardsPositive	Round the result towards $+\infty$.
roundTiesToAway	Round the result to the closest representable value. Round ties to the larger in magnitude.
roundTiesToEven	Round the result to the closest representable value. Round ties towards the even value.
roundTowardsZero	Round the result towards zero.
roundTowardsNegative	Round the result towards $-\infty$.

with the default action to *roundTiesToEven*.

Storage Order. For the examples in this section we have implicitly assumed that the bytes of the DFP number are stored in memory with the most significant byte first. This is called big-endian format and matches the way the bits are labeled in the diagrams of the standard. Naturally, the concept of big-endianness is more general than just this one instance. The library we will be using later in the chapter uses little-endian format. The difference is simply reversing the order of the bytes in memory. For convenience, our examples will continue to use big-endian format.

An Example. Before we move on and start using the open source decimal floating-point library let's use what we've covered in this section to write a simple routine to parse a DFP number, stored as eight bytes, and output the value as an ASCII string. We will assume the storage order to be big-endian.

The main function, dfp_parse() is listed in Fig. 6.1. It accepts a big-endian *decimal64* number as an array of bytes and fills in a given string with the output value. The string array is assumed to be large enough.

In line 4 the sign of the output string is set. If the number is not negative the sign will be overwritten. Lines 5 through 7 extract the sign bit, *CF* field and *BXCF* field. The *BXCF* field is made up of the last two bits of the first byte and all of the second. At this point we can do two quick checks to see if the number is infinity or a NaN. If so, copy the proper string to the output and return (lines 9 and 10). Line 12 calls a helper function, dfp_parse_exp(), to extract the exponent and first output digit from the combined *CF* and *BXCF* fields. This function is,

```
 1 |int dfp_parse_exp(int cf, int bxcf, int *d) {
 2 |    static int digits[] = {0,1,2,3,4,5,6,7,
 3 |                           0,1,2,3,4,5,6,7,
 4 |                           0,1,2,3,4,5,6,7,
 5 |                           8,9,8,9,8,9};
 6 |    static int bits[] = {0,0,0,0,0,0,0,0,
 7 |                         1,1,1,1,1,1,1,1,
 8 |                         2,2,2,2,2,2,2,2,
 9 |                         0,0,1,1,2,2};
10 |
11 |    *d = digits[cf];
12 |    return ((bits[cf] << 8) + bxcf) - 398;
13 |}
```

which is an encoding of Table 6.1 so that the 5-bit *CF* value can be used to look up the proper digit and bits. The bits are combined with the *BXCF* to produce the biased exponent which is then reduced by subtracting the bias term (line 12), which is returned.

Once we have the exponent and first significand digit we can put it in the output (line 14) after setting up our index (line 13) which will cause the initial minus sign to be overwritten if the number is positive. Lines 16 through 40 appear daunting at first but they are really five repetitions of the same pattern. First we extract the 10 bits of a declet (lines 16, 21, 26, 31, and 36). Next, we recover the three decimal digits from this declet by calling the helper function dfp_parse_declet(). We then place the digits in the output string. The lines defining the declet are somewhat cryptic. They

```
 1  void dfp_parse(unsigned char *b, unsigned char *s) {
 2      int sign, cf, bxcf, d0,d1,d2, dec, exponent, i=0;
 3
 4      s[0] = '-';
 5      sign = (b[0] & 0x80) != 0;
 6      cf = (b[0] >> 2) & 0x1f;
 7      bxcf = ((b[0] & 0x3) << 6) + (b[1] >> 2);
 8
 9      if (cf==0x1e) { strcpy(&s[sign],"inf"); return; }
10      if (cf==0x1f) { strcpy(s,"NaN"); return; }
11
12      exponent = dfp_parse_exp(cf, bxcf, &d0);
13      i += sign;
14      s[i++] = d0 + '0';
15
16      dec = ((b[1] & 0x3) << 8) + b[2];
17      dfp_parse_declet(dec, &d0, &d1, &d2);
18      s[i++] = d0 + '0';
19      s[i++] = d1 + '0';
20      s[i++] = d2 + '0';
21      dec = (b[3] << 2) + ((b[4] >> 6) & 0x3);
22      dfp_parse_declet(dec, &d0, &d1, &d2);
23      s[i++] = d0 + '0';
24      s[i++] = d1 + '0';
25      s[i++] = d2 + '0';
26      dec = ((b[4] & 0x3f) << 4) + ((b[5] & 0xf0) >> 4);
27      dfp_parse_declet(dec, &d0, &d1, &d2);
28      s[i++] = d0 + '0';
29      s[i++] = d1 + '0';
30      s[i++] = d2 + '0';
31      dec = ((b[5] & 0xf) << 6) + ((b[6] & 0xfc) >> 2);
32      dfp_parse_declet(dec, &d0, &d1, &d2);
33      s[i++] = d0 + '0';
34      s[i++] = d1 + '0';
35      s[i++] = d2 + '0';
36      dec = ((b[6] & 0x3) << 8) + b[7];
37      dfp_parse_declet(dec, &d0, &d1, &d2);
38      s[i++] = d0 + '0';
39      s[i++] = d1 + '0';
40      s[i++] = d2 + '0';
41
42      s[i++] = 'E';
43      sprintf(&s[i], "%d", exponent);
44  }
```

Fig. 6.1 A function to parse a *decimal64* number

are pulling the 10 bits of the declet from the bytes of the input DFP number using
standard bit manipulation to piece the declet together. A helpful exercise would be to
write out all the bits for one of the hexadecimal examples above, say the one for 0.1,
and then mark off the five declets. This will make clear the origin of the particular

incantations used in this function. The helper function itself is an implementation of Table 6.2 and is given in Fig. 6.2 with `dfp_parse_bits()` defined as,

```
1 int dfp_parse_bits(int a, int b, int c, int d) {
2     return 8*a + 4*b + 2*c + d;
3 }
```

The helper function assigns the bits from the input declet (lines 4 through 8) and then compares them to the values from the table in lines 10 through 44. When there is a match, the three decimal numbers are extracted by setting the proper bits according to Table 6.2.

After line 40 in Fig. 6.1 all that remains is to copy the exponent to the output string which is done in lines 42 and 43.

This example is only the simplest of ways to convert the DFP number to a decimal string. The output is given without any attempt to remove leading zeros of the significand or to scale the number to a more standard range. For example, if the function `dfp_parse()` is given 22 18 00 00 19 43 65 65 as input (in hexadecimal) it produces

0000000314159265E-8

as output. Clearly, a more complete implementation would change this to something closer to the more familiar 3.14159265 format. We leave implementing a function for *decimal32* DFP numbers to the exercises.

6.3 Decimal Floating-Point in Software

In this section we look at DFP implementations in C and Python along with some examples of their use. DFP operations are complex enough that beyond the example above to parse a DFP number it does not make sense to attempt an implementation of our own in this case.

DFP in C. DFP implementations in hardware are not yet common. However, IBM has released, as open source, a DFP library which is compatible with the IEEE 754 standard. It is this library that we will be using in this section. The *decNumber* library is available here,

http://speleotrove.com/decimal/decnumber.html

along with some documentation on its use. We are using version 3.68 but later versions should operate in much the same way. Download the library and expand it. There is no Makefile or installation, the library is simply a set of source code files in C along with documentation and a few examples. The IEEE compliant portion is implemented in the `decimal32.c`, `decimal64.c` and `decimal128.c` files. We will focus on the *decimal64* implementation. Explore the other options the library provides as we will only scratch the surface here.

```
1  void dfp_parse_declet(int dec, int *d0, int *d1, int *d2) {
2      int p,q,r,s,t,u,v,w,x,y;
3
4      y = (dec & 0x1) != 0;    x = (dec & 0x2) != 0;
5      w = (dec & 0x4) != 0;    v = (dec & 0x8) != 0;
6      u = (dec & 0x10) != 0;   t = (dec & 0x20) != 0;
7      s = (dec & 0x40) != 0;   r = (dec & 0x80) != 0;
8      q = (dec & 0x100) != 0;  p = (dec & 0x200) != 0;
9
10     if (v==0) {
11         *d0 = dfp_parse_bits(0,p,q,r);
12         *d1 = dfp_parse_bits(0,s,t,u);
13         *d2 = dfp_parse_bits(0,w,x,y);
14     }
15     if ((v==1) && (x==0) && (w==0)) {
16         *d0 = dfp_parse_bits(0,p,q,r);
17         *d1 = dfp_parse_bits(0,s,t,u);
18         *d2 = dfp_parse_bits(1,0,0,y);
19     }
20     if ((v==1) && (x==0) && (w==1)) {
21         *d0 = dfp_parse_bits(0,p,q,r);
22         *d1 = dfp_parse_bits(1,0,0,u);
23         *d2 = dfp_parse_bits(0,s,t,y);
24     }
25     if ((v==1) && (x==1) && (w==0)) {
26         *d0 = dfp_parse_bits(1,0,0,r);
27         *d1 = dfp_parse_bits(0,s,t,u);
28         *d2 = dfp_parse_bits(0,p,q,y);
29     }
30     if ((v==1) && (x==1) && (w==1) && (s==0) && (t==0)) {
31         *d0 = dfp_parse_bits(1,0,0,r);
32         *d1 = dfp_parse_bits(0,p,q,u);
33         *d2 = dfp_parse_bits(1,0,0,y);
34     }
35     if ((v==1) && (x==1) && (w==1) && (s==1) && (t==0)) {
36         *d0 = dfp_parse_bits(0,p,q,r);
37         *d1 = dfp_parse_bits(1,0,0,u);
38         *d2 = dfp_parse_bits(1,0,0,y);
39     }
40     if ((v==1) && (x==1) && (w==1) && (s==1) && (t==1)) {
41         *d0 = dfp_parse_bits(1,0,0,r);
42         *d1 = dfp_parse_bits(1,0,0,u);
43         *d2 = dfp_parse_bits(1,0,0,y);
44     }
45
46     return;
47  }
```

Fig. 6.2 A function to parse a DPD declet into three BCD digits

The library uses little-endian storage order. A useful example is `example5.c` which we can trivially modify to output in big-endian order. In a text editor, open `example5.c` and change line 30 from,

```
sprintf(&hexes[i*3], "%02x ", a.bytes[i]);
```

to,

```
sprintf(&hexes[i*3], "%02x ", a.bytes[7-i]);
```

so that bytes are output with the highest order byte first. With this change the program will output hex digits as we have been using them above. Compile the program with,

```
gcc example5.c -o example5 decimal64.c decNumber.c decContext.c
```

then test the program with,

```
./example5 0.1
```

which should produce,

```
0.1 => 22 34 00 00 00 00 00 01 => 0.1
```

matching the hexadecimal numbers we decoded in the previous section. This confirms that the *decNumber* library is working properly.

The full *decNumber* library is far more complete than what we will need here. We will work with IEEE 754 compliant functions as defined in the `decDouble.c` file. These functions ensure that we get results matching IEEE 754 decimal floating-point hardware. In order to compare results and performance between DFP and binary floating-point we will again make use of the logistic map which we first saw in Chap. 3. To review, this map iterates values between $[0, 1]$ and can be used to demonstrate the bifurcation route to chaotic dynamic behavior, among other things. For the present, it is sufficient to know that this map has the form,

$$x_{i+1} = rx_i(1 - x_i)$$

for some initial x_0, which we set to 0.25 and a fixed r of 3.8 which is in the chaotic region.

First we will implement this map in C using variables of type `double`. We will then implement the map using the *decNumber* library and compare outputs and execution time. This will also illustrate how to use the library functions to set up a context, initialize variables, and perform arithmetic.

Figure 6.3 shows the plain C version. This version will initialize x and then output the first eight iterates. It then runs for 10 million more iterates and outputs the final eight. Note that we have broken the calculation into individual operations (lines 9 to 11) so that the DFP implementation matches in terms of the order in which the expressions are evaluated.

```
 1  #include <stdio.h>
 2
 3  int main(int argc , char *argv[]) {
 4      double a,b;
 5      double x=0.25, r=3.8;
 6      int i;
 7
 8      for(i=0; i < 8; i++) {
 9          a = 1.0 - x;
10          b = x * a;
11          x = r * b;
12          printf("%0.16f\n", x);
13      }
14
15      for(i=0; i < 10000000; i++) {
16          a = 1.0 - x;
17          b = x * a;
18          x = r * b;
19      }
20
21      printf("\n");
22
23      for(i=0; i < 8; i++) {
24          a = 1.0 - x;
25          b = x * a;
26          x = r * b;
27          printf("%0.16f\n", x);
28      }
29
30      return 0;
31  }
```

Fig. 6.3 A program to iterate the logistic map, $x_{i+1} = rx_i(1 - x_i)$, using IEEE 754 binary floating-point

Figure 6.4 shows the *decNumber* library version. This requires some explanation. Before the main program we define a helper function, pp(), which will print a *double64* number by first converting it to a string. The function uses the decDoubleToString() library function. The constant DECDOUBLE_String is large enough to hold the biggest string that might be produced. The argument is a pointer to a decDouble. This is the base type for a *decimal64* number. The type is a structure which does not use any heap memory which means that it can be created, used, and ignored like a normal C variable.

The program starts in line 12 by defining the temporary variables we will use to store the logistic map calculation. Line 13 declares a context. This will be initialized to properly handle a *decimal64* number in line 17. This context must be passed to many of the library functions to get the expected behavior. The context includes the rounding mode. Here we leave it to the default. We will describe how to work with rounding modes below. Line 14 defines our variable, *x*, as well as *r* and a constant *one*. Initial values are set from strings in lines 19 through 21. Since this is base ten, the value set is exact. Lines 23 through 28 iterate the equation eight times printing the value of x after each iteration. These will be compared to the initial values generated by the binary floating-point version. The equation itself is implemented in a way that mirrors the binary implementation but using the decDoubleSubtract() and decDoubleMultiply() library functions. There are similarly named functions for addition and division plus many other operations. The final eight iterates are printed with lines 38 through 43. Compile the programs with,

```
gcc logistic.c -o logistic
gcc logistic64.c -o logistic64 decContext.c decDouble.c decQuad.c
```

where logistic64.c needs both decDouble.c and decQuad.c so that computations can be performed with 128-bits and then rounded to 64-bits.

Let's compare the output of the two programs, Figs 6.3 and 6.4. The first eight iterates for logistic and logistic64 are,

logistic	logistic64
0.7124999999999999	0.71250
0.7784062500000002	0.77840625000
0.6554618478515620	0.6554618478515625
0.8581601326777955	0.8581601326777949
0.4625410135688510	0.4625410135688527
0.9446679324750937	0.9446679324750942
0.1986276333476368	0.1986276333476352
0.6048638471497435	0.6048638471497399

where we already see that after a few iterations the two sequences begin to diverge. Note that the decimal result is exact at least for the first two iterations while the binary result is already showing rounding error. The final eight iterations are,

logistic	logistic64
0.6245962562304587	0.9494112872063385
0.8910079371467448	0.1825120807398593
0.3690286137353397	0.5669654002706923
0.8848166847236699	0.9329594136330268
0.3872812528014777	0.2376753553568733
0.9017190393139833	0.6885059670888577
0.3367628911200712	0.8149689014131991
0.8487438558811845	0.5730194463417375

where there is no longer any relationship between the two sequences. This is after 10,000,000 iterations. In this case, the departure is not fatal due to the shadowing theorem which states that for any chaotic trajectory the numerically computed trajectory using that initial condition (here 0.25) ultimately follows another chaotic trajectory with a slightly different initial condition. This theorem means that chaotic trajectories generated by computers are real.

```
1  #include "decDouble.h"
2  #include <stdio.h>
3
4  void pp(decDouble *x) {
5      char s[DECDOUBLE_String];
6
7      decDoubleToString(x, s);
8      printf("%s", s);
9  }
10
11 int main(int argc , char *argv[]) {
12     decDouble a,b;
13     decContext set;
14     decDouble x,r,one;
15     int i;
16
17     decContextDefault(&set, DEC_INIT_DECDOUBLE);
18
19     decDoubleFromString(&x,   "0.25", &set);
20     decDoubleFromString(&r,    "3.8", &set);
21     decDoubleFromString(&one,    "1", &set);
22
23     for(i=0; i < 8; i++) {
24         decDoubleSubtract(&a, &one, &x, &set);
25         decDoubleMultiply(&b, &x, &a, &set);
26         decDoubleMultiply(&x, &r, &b, &set);
27         pp(&x); printf("\n");
28     }
29
30     for(i=0; i < 10000000; i++) {
31         decDoubleSubtract(&a, &one, &x, &set);
32         decDoubleMultiply(&b, &x, &a, &set);
33         decDoubleMultiply(&x, &r, &b, &set);
34     }
35
36     printf("\n");
37
38     for(i=0; i < 8; i++) {
39         decDoubleSubtract(&a, &one, &x, &set);
40         decDoubleMultiply(&b, &x, &a, &set);
41         decDoubleMultiply(&x, &r, &b, &set);
42         pp(&x); printf("\n");
43     }
44
45     return 0;
46 }
```

Fig. 6.4 A program to iterate the logistic map, $x_{i+1} = rx_i(1-x_i)$, using IEEE 754 decimal floating-point

If we use the Python rational arithmetic class developed in Chap. 4 we see that the first few iterates of the logistic map are exactly,

57/80	= 0.7125000000000000
24909/32000	= 0.7784062500000000
3355964661/5120000000	= 0.6554618478515625
112480764910343946501/	= 0.8581601326777950
13107200000000000000	

which matches the decimal floating-point output more closely than the binary.

Naturally, there is a large difference in performance between logistic and logistic64 since the former is using hardware (on systems using Intel instructions). Logistic64 is approximately 45 times slower than logistic so DFP's precision must be balanced by the runtime hit. For some applications it may not be feasible to use DFP.

We conclude this introduction to the *decNumber* library by looking at rounding modes and their effect on calculations. Rounding modes are set using decContext SetRounding(). A simple example will show the effect. Figure 6.5 illustrates all the IEEE rounding modes using the *decNumber* library. The value is chosen so that the first digit beyond the limit of a *decimal64* number is 5 to enable a tie between representable numbers. The output of Fig. 6.5 is,

Mode	Value
roundTiesToEven	0.1234567890123456
roundTowardsPositive	0.1234567890123457
roundTiesToAway	0.1234567890123457
roundTowardsZero	0.1234567890123456
roundTowardsNegative	0.1234567890123456
roundTiesToEven	-0.1234567890123456
roundTowardsPositive	-0.1234567890123456
roundTiesToAway	-0.1234567890123457
roundTowardsZero	-0.1234567890123456
roundTowardsNegative	-0.1234567890123457

where the last digit of the converted number shows the choice made according to the IEEE rounding rules.

DFP in Python. Python supports decimal floating-point through its decimal module. This module is really a wrapper on the *decNumber* library but is easier to use because of operator overloading. A quick example shows it in action,

```
 1  int main(int argc , char *argv[]) {
 2      decDouble a;
 3      decContext set;
 4
 5      decContextDefault(&set, DEC_INIT_DECDOUBLE);
 6
 7      decContextSetRounding(&set, DEC_ROUND_HALF_EVEN);
 8      decDoubleFromString(&a, "0.12345678901234565", &set);
 9      pp(&a); printf("\n");
10      decContextSetRounding(&set, DEC_ROUND_CEILING);
11      decDoubleFromString(&a, "0.12345678901234565", &set);
12      pp(&a); printf("\n");
13      decContextSetRounding(&set, DEC_ROUND_UP);
14      decDoubleFromString(&a, "0.12345678901234565", &set);
15      pp(&a); printf("\n");
16      decContextSetRounding(&set, DEC_ROUND_DOWN);
17      decDoubleFromString(&a, "0.12345678901234565", &set);
18      pp(&a); printf("\n");
19      decContextSetRounding(&set, DEC_ROUND_FLOOR);
20      decDoubleFromString(&a, "0.12345678901234565", &set);
21      pp(&a); printf("\n\n");
22      decContextSetRounding(&set, DEC_ROUND_HALF_EVEN);
23      decDoubleFromString(&a, "-0.12345678901234565", &set);
24      pp(&a); printf("\n");
25      decContextSetRounding(&set, DEC_ROUND_CEILING);
26      decDoubleFromString(&a, "-0.12345678901234565", &set);
27      pp(&a); printf("\n");
28      decContextSetRounding(&set, DEC_ROUND_UP);
29      decDoubleFromString(&a, "-0.12345678901234565", &set);
30      pp(&a); printf("\n");
31      decContextSetRounding(&set, DEC_ROUND_DOWN);
32      decDoubleFromString(&a, "-0.12345678901234565", &set);
33      pp(&a); printf("\n");
34      decContextSetRounding(&set, DEC_ROUND_FLOOR);
35      decDoubleFromString(&a, "-0.12345678901234565", &set);
36      pp(&a); printf("\n");
37  }
```

Fig. 6.5 A program to illustrate the effect of IEEE 754 decimal rounding modes

```
 1  >>> from decimal import *
 2  >>> "%0.16f" % 0.1
 3      '0.1000000000000000'
 4  >>> "%0.16f" % (1.95-1.85)
 5      '0.0999999999999999'
 6  >>> setcontext(Context(prec=16))
 7  >>> print Decimal("0.1")
 8      0.1
 9  >>> print Decimal("1.95") - Decimal("1.85")
10      0.10
```

After importing the DFP library the first two examples (lines 2 and 4) show why decimal floating-point might be desirable. The output should be the same, 0.1, but it is not because of rounding and the fact that 0.1 is a repeating value in binary.

Line 6 sets the context of the decimal module to use 16 digits. This matches the decDouble library used above and corresponds to an IEEE *decimal64* number. Note that both libraries support arbitrary precision but we restrict ourselves to IEEE sized numbers here.

The Decimal class defines a DFP number, typically by using a string as we did above. Lines 7 and 9 repeat lines 2 and 4 but show that there is no rounding error but only an exact result. Additionally, line 9 returns "0.10" instead of "0.1" to indicate that the result is accurate to two digits.

Figure 6.6 shows a Python implementation of the logistic example given above. The output is identical to that of logstic64 as expected since the underlying library is the same and we have initialized the context to 16 digits of precision. However, the run time is significantly worse, taking 82x longer than the C *decNumber* version. This is due to the Python interpreter itself and the overhead of object creation and not the performance of the C library itself.

```
 1  from decimal import *
 2
 3  def main():
 4      setcontext(Context(prec=16))
 5
 6      x = Decimal("0.25")
 7      r = Decimal("3.8")
 8      one = Decimal("1")
 9
10      for i in xrange(8):
11          a = one - x
12          b = x * a
13          x = r * b
14          print x
15
16      for i in xrange(10000000):
17          a = one - x
18          b = x * a
19          x = r * b
20
21      print
22
23      for i in xrange(8):
24          a = one - x
25          b = x * a
26          x = r * b
27          print x
28
29  main()
```

Fig. 6.6 A Python version of the logistic map using the decimal module

6.4 Thoughts on Decimal Floating-Point

Decimal floating-point has not increased in use as might have been expected. Humans work in base ten almost universally so making calculations in base ten seems quite reasonable. One reason for the slow adoption of DFP is the fact that hardware floating-point in binary is so much faster. The few hardware implementations of DFP are slower than their binary cousins but not terribly so. Still, even a 10% reduction in performance may not be acceptable in some circumstances.

The exactness of the results of DFP calculations makes this a perfect number format for commercial or financial work. Many, if not most, databases are using decimal storage for monetary values, often as character strings in base ten, so that the possible application of DFP to these databases is obvious. It would probably also result in smaller storage requirements and would certainly result in more accuracy or fewer checks for edge cases that cannot be allowed in financial calculations.

Is DFP useful for scientific programming? We think so, if it were fast enough. The logistic example above clearly shows that DFP results are sometimes "truer" than binary floating-point. The preservation of knowledge that there are meaningful trailing zeros in a value is also potentially of significance. The Python example returning "0.10" as the answer instead of "0.1" is helpful as it tells users that the digit in the hundredths place is truly zero and not simply unknown. A scientific result might require a great deal of effort to know that digit is exactly zero so it would be good if the computer representation preserved it.

While writing this chapter two potentially highly useful exercises came to mind. As far as we know as of this writing neither has been done but we present them as a challenge to motivated individuals,

> – Python is widely used in scientific programming because of its well-supported array-processing libraries, particularly NumPy and libraries built on top of it (like SciPy). These array-processing libraries are highly optimized and performant. It would be perhaps quite useful to implement a version of NumPy using decimal floating-point in such a way that if compiled on a system with actual DFP hardware it could take advantage of it automatically.
> – Interval arithmetic adds bounds to numbers by specifying a range in which the actual number exists rather than simply a single value. This requires proper control of rounding modes (see Chap. 7). A decimal floating-point based interval arithmetic package would perhaps be useful and could be combined with a DFP based NumPy for Python.

Hopefully this chapter has illustrated that decimal floating-point is of value and should be of primary concern to developers working with financial data. The thought perhaps should not be "should I use decimal floating-point?" but rather, "Is there any valid reason why I should not use decimal floating-point?"

6.5 Chapter Summary

In this chapter we reviewed the IEEE standard for storing decimal floating-point numbers using the format that seems to be in widest use. We learned how to decode such numbers and produce the corresponding decimal value as a string. We then examined an IEEE compliant C library for decimal floating-point and learned the basics of how to use it. We used the logistic map from chaos theory as an example of how decimal floating-point can be more accurate than binary floating-point for iterated calculations. We then examined a Python core module that uses the C library to implement a decimal floating-point class and showed that it produces the same results as before for the logistic map. We concluded by presenting, as food for future thought, two possible ways in which decimal floating-point might grow in its use which in turn would hopefully lead to more accessible hardware implementations.

Exercises

6.1. Parse the following *decimal64* numbers:
```
29 ff 98 42 d2 e9 64 45
25 ff 18 0c ce ef 26 1f
25 fe 14 44 ee 27 cc 5b
```

6.2. The function `dfp_parse()` in Fig. 6.1 works with *decimal64* numbers only. Create a version that works with *decimal32* numbers. Test it with the following inputs,

$$A2 \; 10 \; C6 \; 15 \rightarrow -3.1415$$
$$F8 \; 00 \; 00 \; 00 \rightarrow -\text{inf}$$
$$22 \; 40 \; 00 \; 01 \rightarrow 0.1$$

6.3. Encode the number 123.456 as a *decimal64* number. First decide on the significand and exponent to use, then encode the exponent and first digit of the significand to create the *CF* and *BXCF* fields. Last, use Table 6.3 to encode the significand mapping the remaining 15 decimal digits to five declets. ***

References

1. IEEE Standards Association. Standard for Floating-Point Arithmetic. IEEE 754-2008 (2008).
2. Duale, A., *et al.* Decimal floating-point in z9: An implementation and testing perspective. IBM Journal of Research and Development 51.1.2 (2007): 217–227.

Chapter 7
Interval Arithmetic

Abstract Floating-point numbers do not map exactly to the real numbers. Also, sometimes there are uncertainties in our knowledge of the true value for a quantity. Both of these situations can be addressed by using interval arithmetic which keeps bounds on the possible value of a number while a calculation is in progress. In this chapter we will describe interval arithmetic, implement basic operations for interval arithmetic in C, discuss functions as they pertain to intervals, examine interval implementation for C and Python, and finally offer some advice on when to use interval arithmetic.

7.1 Defining Intervals

A number like π is exact in the sense that there is no uncertainty as to its numeric value even though in this case it is impossible to write it. However, in the real world, there is almost never complete certainty as to the value of a quantity since most are measurements made by some finite device which has limitations. A scale may only be accurate to the ounce or gram. A ruler may only be accurate to a millimeter. Scientists associate uncertainties with measured quantities as a way to mathematically account for the inexactness of our dealings with reality. Similarly, since computers use a finite number of bits to store floating-point numbers, we know that it is impossible to store just any real number in a computer, we can only store the floating-point number we deem to be "nearest to" the actual real number.

Mathematics has ways to deal with uncertainty, be it through probability and statistics or through error analysis and propagation. The latter are mathematical concepts which involve a function, f, of input variables, say, x and y, and how to determine the uncertainty in $f(x, y)$, denoted as σ_f, given x, y and their uncertainties, σ_x and σ_y. It is the propagation of errors that decides how to calculate σ_f. This is an essential technique for science as a whole and the reader is directed to the very useful presentation in Bevington [1] for examples and advice on how to calculate the

© Springer International Publishing Switzerland 2015
R.T. Kneusel, *Numbers and Computers*, DOI 10.1007/978-3-319-17260-6_7

uncertainty of expressions involving variables with uncertainties. For our purposes we need only consider what σ_x means for a particular x and how to represent that meaning in a computer using floating-point numbers. We will always assume that any floating-point numbers we use are IEEE 754 compliant.

Scientific literature is full of measured quantities which are often given along with some uncertainty as to their actual value. This uncertainty may be estimated, the standard deviation of a set of measurements of a quantity, the mean of a set of measurements of a quantity along with the standard error of the mean, etc. For example, it might be reported that the temperature of a reaction was 74.0 ± 1.3 C where the ± 1.3 portion is the uncertainty. This means that, with some level of confidence, usually 95 %, we know that the temperature of the experiment was in the interval $[72.7, 75.3]$ C because this is 1.3 C less and more than the first value reported. Instead of tracking mean values and uncertainties we could just track the interval itself and work with it. This is the essence of *interval arithmetic* and we will write intervals in the form $t = [72.7, 75.3]$ where the first number is the lower bound on the interval and the second is the upper bound (the bounds are included in the interval). Symbolically we will write $t = [\underline{t}, \bar{t}]$ where $\underline{t}$ is the lower bound and $\bar{t}$ is the upper.

We should slow down a bit here. The previous paragraphs mentioned the concept of "propagation of errors" without describing it and without discussing why one might choose interval arithmetic over propagating uncertainty. Without deriving it, we give the general formula for determining the uncertainty of the value of a function (σ_f) from the uncertainties of the arguments to that function. We assume $f(x, y)$ with argument uncertainties of σ_x and σ_y but the formula holds for any number of arguments, just use a similar term for each additional variable. The formula is,

$$\sigma_f^2 = \sigma_x^2 \left(\frac{\partial f}{\partial x} \right)^2 + \sigma_y^2 \left(\frac{\partial f}{\partial y} \right)^2$$

under the assumption that x and y are independent (i.e., not correlated with each other so that the covariance terms disappear) and that x and y represent random variables from Gaussian distributions.

Do not be concerned if the last sentence is not clear. The point to remember is that the formula used for propagation of errors is based on assumptions about the way in which the values behave, or, rather, the environment in which they were produced. And, as is always the case, an assumption may be incorrect. With intervals there is no assumption beyond the statement that the true value lies within the interval.

We will see below how to calculate intervals but let's quickly look at an example comparing intervals to propagation of errors. In this example we simply want to calculate $f(x, y) = x + y$ and find the uncertainty or interval containing $f(x, y)$. We use a C program which uses interval arithmetic to add x and y and to calculate σ_f from the formula above. When we do this we get the following answers,

propagation of errors $= [23.344, 23.564]$
interval arithmetic $= [23.337, 23.571]$

for $x = [12.12, 12.34]$ and $y = [11.217, 11.231]$. Notice that while not exact the two methods produce comparable answers so we know at least that interval arithmetic might be a useful way to account for uncertainty.

In some sense the origin of interval arithmetic stretches back into antiquity. Archimedes, in his *On the measurement of the circle* estimated the value of π by using a two-sided bound which is in effect an interval since he showed that $\underline{x} \leq \pi \leq \bar{x}$ for the bounds $[\underline{x}, \bar{x}]$ which he placed on the value of π.

In modern times intervals were discussed by Young [2] (1931) and extended by Dwyer [3] (1951). Sunaga again extended the algebra of intervals [4] (1958). Dwyer discussed computing with approximate numbers but it was Moore who first wrote about implementing interval arithmetic on digital computers [5] (1959). Moore refers to Dwyer and then extends intervals to floating-point numbers representable on the computer. Indeed, [5] is more than simply interval arithmetic and is an interesting historical work which discusses how to represent floating-point numbers on digital computers. Note that Moore calls interval arithmetic "range arithmetic" after Dwyer and then extends this to "digital range arithmetic" by which is meant the interval arithmetic that is the subject of this chapter.

The remainder of this chapter discusses interval arithmetic over basic operations $(+, -, \times, /,$ comparison$)$ with examples in C and Python. Next comes a discussion of functions and intervals. This includes elementary functions like sine, cosine, and exponential along with general functions made up of expressions involving variables which are themselves intervals. We will see that general functions lead to some undesirable consequences and discuss the "dependency effect" which is the biggest drawback of interval arithmetic. After this we look at intervals in programming, examining some implementations. We conclude the chapter with thoughts on intervals and their use.

7.2 Basic Operations

Basic arithmetic operations are well-suited to working with intervals but before we start our library of functions we need to consider carefully how we will go about defining intervals using floating-point numbers. Mathematically, the bounds of an interval are real numbers and as such are exact. Unfortunately, computers do not operate on real numbers but instead use floating-point numbers which are by necessity a finite set. So, when an operation produces a result, and this is true even when the operands are already floating-point numbers, the result is seldom a floating-point number but must be rounded in some way to be expressed as a floating-point number. How should the result be rounded? IEEE 754 floating-point, which is the standard, has multiple rounding modes. By default, IEEE 754 states that results should be rounded to the nearest floating-point value. This is a completely reasonable thing to do but completely wrong when it comes to interval arithmetic.

If we write our interval arithmetic functions so that they round results to the nearest floating-point number we will quickly see that this introduces a bias that will make the intervals change so that we can no longer be certain that the actual number lies in the specified interval. We can fix this issue, however, by making use of the non-default rounding modes supported by IEEE 754. One of these modes is to always round down towards $-\infty$. Another is to round up towards $+\infty$. These two modes are what we need. When computing the lower bound of an interval we will always round towards $-\infty$ while for an upper bound we will round towards $+\infty$ (see [6]). This will ensure that our interval does not shrink during a calculation.

We begin our library by including required C libraries and by defining the data structure we will use to represent an interval,

```
1  #include <fenv.h>
2  #include <stdio.h>
3  #include <math.h>
4
5  typedef struct {
6      double a;
7      double b;
8  } interval_t;
9
10 char *pp(interval_t x) {
11     static char s[100];
12     sprintf(s, "[%0.16g, %0.16g]", x.a, x.b);
13     return s;
14 }
15
16 interval_t interval(double a, double b) {
17     interval_t ans;
18     ans.a = a;
19     ans.b = b;
20     return ans;
21 }
```

where we will use the function `pp()` to display an interval which is of type `interval_t`. To maximize precision we use C `doubles` with a the lower bound and b the upper. The function `interval()` builds an interval from two floating-point numbers representing the bounds.

Note the inclusion of not only `stdio.h` and `math.h` but also `fenv.h`. This latter include is the floating-point library and it defines constants and functions related to floating-point including those necessary to change the rounding mode. Including `fenv.h` is not enough on its own, however. We must also compile our programs with the `-frounding-math` compiler switch (for `gcc`) otherwise the function calls for changing the rounding mode will be ignored.

Addition and Subtraction. Now, let's take a look at how to add two intervals $x = [\underline{x}, \bar{x}]$ and $y = [\underline{y}, \bar{y}]$. In this case, the answer is not too surprising,

$$x + y = [\underline{x}, \bar{x}] + [\underline{y}, \bar{y}] \equiv [\underline{x} + \underline{y}, \bar{x} + \bar{y}]$$

where we must remember in our implementation to use rounding towards $-\infty$ for the lower bound and rounding towards $+\infty$ for the upper. Intuitively it makes sense

that the lower bound would simply be the sum of the two lower bounds and likewise
for the upper bounds. So far, so good. Here's the C code,

```
 1  interval_t int_add(interval_t x, interval_t y) {
 2      interval_t sum;
 3
 4      fesetround(FE_DOWNWARD);
 5      sum.a = x.a + y.a;
 6      fesetround(FE_UPWARD);
 7      sum.b = x.b + y.b;
 8      fesetround(FE_TONEAREST);
 9      return sum;
10  }
```

where lines 5 and 7 add the lower and upper bounds. Line 4 sets the floating-point
rounding mode towards $-\infty$ just before adding the lower bounds. Line 6 then sets the
rounding mode towards $+\infty$ before adding the upper bounds in line 7. Lastly, line 8
returns to rounding towards the nearest floating-point number just before returning
the new sum in line 9.

Subtraction is, as might be expected, similar in form to addition but with a twist.
Interval subtraction is defined to be,

$$x - y = [\underline{x}, \bar{x}] - [\underline{y}, \bar{y}] \equiv [\underline{x} - \bar{y}, \bar{x} - \underline{y}]$$

which at first glance might seem a bit odd. One way to see why subtraction is defined
like this is to consider what negation of an interval might mean. To negate a value is
to change its sign, $x \rightarrow -x$. For an interval the operation is the same but not only are
the signs of the bounds changed but also their order. If the order were not changed
the lower bound would suddenly be larger than the upper bound. So, negation is,

$$-[\underline{x}, \bar{x}] \equiv [-\bar{x}, -\underline{x}]$$

where we have swapped the order of the bounds. How does this help in understanding subtraction? By remembering that subtraction is simply addition of a negative
value. So,

$$[\underline{x}, \bar{x}] - [\underline{y}, \bar{y}] = [\underline{x}, \bar{x}] + [-\bar{y}, -\underline{y}] \equiv [\underline{x} - \bar{y}, \bar{x} - \underline{y}]$$

which is exactly the definition of interval subtraction. In code this looks like,

```
 1  interval_t int_sub(interval_t x, interval_t y) {
 2      interval_t diff;
 3
 4      fesetround(FE_DOWNWARD);
 5      diff.a = x.a - y.b;
 6      fesetround(FE_UPWARD);
 7      diff.b = x.b - y.a;
 8      fesetround(FE_TONEAREST);
 9      return diff;
10  }
```

where we set the proper rounding mode as in addition and lines 5 and 7 set the lower
and upper bounds using the definition of interval subtraction.

Multiplication. The goal of interval arithmetic is to ensure that the interval represents the entire range in which the actual value may be found. This means that
the lower bound must be the smallest such value and the upper bound must be the
largest. For the addition of two intervals it is clear that the lower bound must be simply the sum of the lower bounds of the operands and similarly for the upper bound.
With subtraction the same holds after we see that it is addition of the negation of the
second operand. For multiplication the situation is more general, however, because
the interval may include bounds of opposite signs. Therefore, we define multiplication of two intervals as,

$$[\underline{x}, \bar{x}] \times [\underline{y}, \bar{y}] \equiv [\min\{\underline{xy}, \underline{x}\bar{y}, \bar{x}\underline{y}, \bar{x}\bar{y}\}, \max\{\underline{xy}, \underline{x}\bar{y}, \bar{x}\underline{y}, \bar{x}\bar{y}\}]$$

which reduces to,

$$[\underline{x}, \bar{x}] \times [\underline{y}, \bar{y}] = [\underline{xy}, \bar{x}\bar{y}]$$

if the lower bounds of x and y are greater than zero. This definition covers all cases
in order to find the smallest and largest product of the components of x and y.

With multiplication defined we can now implement it in our library,

```
 1 interval_t int_mul(interval_t x, interval_t y) {
 2     interval_t prod;
 3     double t;
 4
 5     fesetround(FE_DOWNWARD);
 6     prod.a = x.a * y.a;
 7     t = x.a * y.b;
 8     if (t < prod.a) prod.a = t;
 9     t = x.b * y.a;
10     if (t < prod.a) prod.a = t;
11     t = x.b * y.b;
12     if (t < prod.a) prod.a = t;
13
14     fesetround(FE_UPWARD);
15     prod.b = x.a * y.a;
16     t = x.a * y.b;
17     if (t > prod.b) prod.b = t;
18     t = x.b * y.a;
19     if (t > prod.b) prod.b = t;
20     t = x.b * y.b;
21     if (t > prod.b) prod.b = t;
22
23     fesetround(FE_TONEAREST);
24     return prod;
25 }
```

where we accept two input intervals in x and y and return prod as their product.
Lines 5 through 12 set the lower bound so we first set the IEEE rounding mode to
$-\infty$ (line 5). Line 6 sets the lower bound of prod to $\underline{xy}$. This will be our lower bound

for the product unless one of the other component products is smaller. We set the auxiliary variable t to the next candidate (line 7) and in line 8 check to see if this is smaller than our current lower bound. If so, we update the lower bound. Lines 9 through 12 continue this process for the remaining component products so that after line 12 we know prod.a is set to the smallest value. Line 14 sets the rounding mode to $+\infty$ and repeats this process for the upper bound keeping the largest component product in prod.b. Line 23 then restores the default IEEE rounding mode and the final product interval is returned in line 24.

Reciprocal and Division. Just as subtraction is really the addition of a negative value so it is that division is really multiplication by the reciprocal. Therefore, in order to discuss division, we need to take a look at reciprocal in interval arithmetic. Our first thought would be to simply take the reciprocal, $x \to 1/x$, of each component. However, just as negation of an interval required swapping the bounds so will the reciprocal because $1/\bar{x} < 1/\underline{x}$, like so,

$$1/[\underline{x}, \bar{x}] \equiv [1/\bar{x}, 1/\underline{x}]$$

where we have swapped the bounds as needed. However, there is an important caveat that we must keep in mind. Division by zero is mathematically undefined so we must be careful of the case when the interval we are taking the reciprocal of contains zero. If the interval does not contain zero, which means that $\underline{x} > 0$ or $\bar{x} < 0$, we can proceed normally because the reciprocal is defined over the entire interval. This means we must add a qualifier to our original definition,

$$1/[\underline{x}, \bar{x}] \equiv [1/\bar{x}, 1/\underline{x}], \underline{x} > 0 \text{ or } \bar{x} < 0$$

which makes it explicit that we are not allowed to consider intervals which contain zero.

For scalars, we simply say that division by zero is undefined and leave it at that. Can we say something more in the case of intervals? One approach is to split the interval into two parts at zero,

$$1/[\underline{x}, 0] \to [-\infty, 1/\underline{x}]$$
$$1/[0, \bar{x}] \to [1/\bar{x}, \infty]$$

following the definition above and writing $\pm\infty$ for $1/0$. If we do this we get two overlapping intervals representing the reciprocal of our original interval, x, which contained zero. Because of this we are really saying that there is no information in writing the reciprocal of an interval containing zero, but we could continue to work through an expression involving such an interval if we use both the intervals above. This is one reason why interval arithmetic is not more widely used, though it isn't the main one (we will see that below). For our convenience, and because we are trying to be pedagogical and not obfuscate things, we will declare reciprocals of intervals which contain zero to be illegal and ignore them.

Armed with a definition of the reciprocal of an interval we are now ready to calculate x/y. We do this by first calculating the reciprocal, $y' \equiv 1/y$, and then find $x \times y'$. The C code we need is,

```
 1 interval_t int_recip(interval_t x) {
 2     interval_t inv;
 3
 4     if ((x.a < 0.0) && (x.b > 0.0)) {
 5         inv.a = -exp(1000.0);
 6         inv.b =  exp(1000.0);
 7     } else {
 8         fesetround(FE_DOWNWARD);
 9         inv.a = 1.0 / x.b;
10         fesetround(FE_UPWARD);
11         inv.b = 1.0 / x.a;
12         fesetround(FE_TONEAREST);
13     }
14
15     return inv;
16 }
17
18 interval_t int_div(interval_t x, interval_t y) {
19     return int_mul(x,int_recip(y));
20 }
```

where we call the `int_mul()` function defined above to do the actual multiplication with proper rounding modes.

The function `int_div()` is straightforward, simply multiply the first argument by the reciprocal of the second as in line 19 above. All the action is in `int_recip()` which first checks to see if zero is in the interval (line 4). If zero is in the interval we call the reciprocal undefined and return $[-\infty, +\infty]$ by attempting to calculate the exponential of a large number (lines 5 and 6). Line 8 sets rounding towards $-\infty$ and then calculates the lower bound in line 9. This is repeated by lines 10 and 11 with rounding set towards $+\infty$. Line 12 resets the rounding mode and line 15 returns the new interval.

Powers. What it means to raise a number to a positive integer power is clear, simply multiply the number by itself that many times so that $x^4 = x \times x \times x \times x$. For intervals we must be more careful. The result of raising an interval to an integer power depends both upon the even or oddness of the exponent and the signs of the components of the interval.

Let's present the formulas and then look at why they work. For an odd exponent, $n = 2k + 1$, $k = 0, 1, \ldots$, things work as expected,

$$[\underline{x}, \bar{x}]^n = [\underline{x}^n, \bar{x}^n], \quad n = 2k+1, \quad k = 0, 1, \ldots$$

It is only when the exponent is even that things become interesting. In this case we have,

$$[\underline{x},x]^n = [\underline{x}^n, \bar{x}^n], \quad \underline{x} \geq 0, \quad n = 2k, \quad k = 1, 2, \ldots$$
$$= [\bar{x}^n, \underline{x}^n], \quad \bar{x} < 0$$
$$= [0, \max\{\underline{x}^n, \bar{x}^n\}], \quad \text{otherwise}$$

For an odd exponent it is straightforward to see why the formula takes that form. If the interval has bounds that are negative then an odd power of those bounds will still be negative and the order of the original will not change. This means that for an odd exponent $\underline{x}^n < \bar{x}^n$ so the answer is simple, no special cases. When the exponent is even, however, we have to pay attention to the signs of the bounds of x. In the first case, if $\underline{x} \geq 0$ we know that both bounds are positive and that $\underline{x} < \bar{x}$. Therefore, the order of the bounds need not change since $\underline{x}^n < \bar{x}^n$ as well. The second case tells us that both bounds are negative. Since $a^n = |a|^n$ if n is even and $a < 0$ it follows that $\underline{x}^n > \bar{x}^n$ when $\underline{x}, \bar{x} < 0$ so we must swap the order of the bounds. The last case covers situation when the interval includes zero. As x^n is always positive when n is even (and a positive integer) we know that the lower bound of the result must be zero. The upper bound is set to the larger of the two original bounds after raising them to the n-th power.

What if n is a negative integer? In that case we use the rule that $x^{-n} = 1/x^n$. To do this, first calculate $y = x^n = [\underline{x}, \bar{x}]^n$ and then calculate the result as $[\underline{x}, \bar{x}]^{-n} = [1/\bar{y}, 1/\underline{y}]$ with suitable warnings about y containing zero.

Adding int_pow() to our library is straightforward, if a little tedious in checking all the special cases. The C code is,

```
 1  interval_t int_pow(interval_t x, int n) {
 2      interval_t ans;
 3
 4      if ((n%2) == 1) {
 5          fesetround(FE_DOWNWARD);
 6          ans.a = pow(x.a,n);
 7          fesetround(FE_UPWARD);
 8          ans.b = pow(x.b,n);
 9      } else {
10          if (x.a >= 0) {
11              fesetround(FE_DOWNWARD);
12              ans.a = pow(x.a,n);
13              fesetround(FE_UPWARD);
14              ans.b = pow(x.b,n);
15          } else {
16              if (x.b < 0) {
17                  fesetround(FE_DOWNWARD);
18                  ans.a = pow(x.b,n);
19                  fesetround(FE_UPWARD);
20                  ans.b = pow(x.a,n);
21              } else {
22                  ans.a = 0.0;
23                  fesetround(FE_UPWARD);
```

```
24│                              ans.b = pow(x.a,n);
25│                              if (pow(x.b,n) > ans.b) ans.b = pow(x.b,n);
26│                         }
27│                    }
28│               }
29│
30│          fesetround(FE_TONEAREST);
31│          return ans;
32│}
```

where line 4 checks to see if the exponent n is even or odd. If the remainder after dividing by 2 is 1, then n is odd and we calculate the lower bound in line 6 and the upper in line 8 after setting the floating-point rounding mode appropriately. If n is even we move to line 10 and check the three cases given above. If the lower bound is greater than or equal to zero (line 10) we calculate the new bounds in lines 12 and 14. If the upper bound is less than zero (line 16) we swap the results (lines 18 and 20). Finally, if the interval includes zero we set the lower bound to zero (line 22) and set the upper bound to the larger of the two exponentiated initial bounds (lines 24 and 25). Line 30 restores the default rounding mode and line 31 returns our answer.

When using intervals it is important to remember that, unlike real numbers, exponentiation is not simply repeated multiplication. Let's look at some examples using our library routines above. We will give an example of an even and odd power of intervals that are positive, negative, and include zero. We will use int_pow() for the powers and int_mul() for the multiplications.

First, an all negative interval,

$$
\begin{aligned}
x &= [\ -2.7182818284590451,\ -2.6182818284590454] \\
x^3 &= [-20.0855369231876679,\ -17.9493685483622478] \\
x \times x \times x &= [-20.0855369231876679,\ -17.9493685483622443] \\
x^4 &= [\ 46.9965055024911891,\ 54.5981500331442291] \\
x \times x \times x \times x &= [\ 46.9965055024911749,\ 54.5981500331442433]
\end{aligned}
$$

Second, an all positive interval,

$$
\begin{aligned}
x &= [\ 2.7182818284590451,\ 2.8182818284590452] \\
x^3 &= [20.0855369231876644,\ 22.3848022077206359] \\
x \times x \times x &= [20.0855369231876608,\ 22.3848022077206359] \\
x^4 &= [54.5981500331442220,\ 63.0866812956689813] \\
x \times x \times x \times x &= [54.5981500331442149,\ 63.0866812956689884]
\end{aligned}
$$

Lastly, an interval including zero,

$$
\begin{aligned}
x &= [\ -2.718281828459051,\ 2.8182818284590452] \\
x^3 &= [-20.0855369231876679,\ 22.3848022077206359] \\
x \times x \times x &= [-21.5905309612583913,\ 22.3848022077206359] \\
x^4 &= [\ 0.0000000000000000,\ 63.0866812956689813] \\
x \times x \times x \times x &= [-60.8482010748969273,\ 63.0866812956689884]
\end{aligned}
$$

The examples above are displayed using a format specifier of %0.16f which always displays 16 digits after the decimal point. Note that the repeated multiplications tend to lead to result intervals that are a little too large compared to the powers. Especially note that the even power of the interval including zero is completely wrong in the sense that it is far too large by including a negative lower bound.

Other Operations. There are many other operations we could add to our library. Here we add only a few of these. The new operations are negation, absolute value, comparison for equality, comparison for less than or equal, and a concept of "distance" between two intervals. We will describe the C code first and then give some examples. The code for negation is,

```
1 interval_t int_neg(interval_t x) {
2     interval_t ans;
3     fesetround(FE_DOWNWARD);
4     ans.a = -x.b;
5     fesetround(FE_UPWARD);
6     ans.b = -x.a;
7     fesetround(FE_TONEAREST);
8     return ans;
9 }
```

which implements $-[\underline{x}, \bar{x}] = [-\bar{x}, -\underline{x}]$. This is always true because the lower bound of an interval is, by definition, less than (or equal) to the upper bound. Therefore, changing the sign of these values means that their order must be reversed to keep the lower bound less than the upper.

The absolute value of an interval depends on the signs of the bounds. Formally, the absolute value is defined to be,

$$|x| = [\min\{|\underline{x}|, |\bar{x}|\}, \max\{|\underline{x}|, |\bar{x}|\}], \text{ if } \underline{x}\bar{x} \geq 0$$
$$= [0, \max\{|\underline{x}|, |\bar{x}|\}], \text{ if } \underline{x}\bar{x} < 0$$

where we maximize the interval if the signs of the bounds are the same ($\underline{x}\bar{x} \geq 0$) or we set zero to be the lower bound if the interval includes zero ($\underline{x}\bar{x} < 0$).

The C code for this operation is,

```
1  interval_t int_abs(interval_t x) {
2      interval_t ans;
3      if (x.b * x.a >= 0) {
4          fesetround(FE_DOWNWARD);
5          ans.a = (fabs(x.b) < fabs(x.a)) ? fabs(x.b) : fabs(x.a);
6          fesetround(FE_UPWARD);
7          ans.b = (fabs(x.b) < fabs(x.a)) ? fabs(x.a) : fabs(x.b);
8      } else {
9          ans.a = 0.0;
10         fesetround(FE_UPWARD);
11         ans.b = (fabs(x.b) < fabs(x.a)) ? fabs(x.a) : fabs(x.b);
12     }
13     fesetround(FE_TONEAREST);
14     return ans;
15 }
```

where line 3 looks to see if the signs of the bounds are the same. If so, the minimum is used for the lower bound (line 5) and the maximum is used for the upper bound (line 7) with proper rounding modes set. If the interval includes zero we move to line 9 and set the output lower bound to zero and the upper bound to the maximum (line 11). We reset to the default rounding mode and return the answer in line 14.

To this point the interval library consists entirely of operations. Here we define the concepts of equality and less than or equal for intervals. Equality is straightforward: two intervals are equal if their bounds are equal. Symbolically this is represented as,

$$[\underline{x}, \bar{x}] = [\underline{y}, \bar{y}] \text{ iff } (\underline{x} = \underline{y}) \wedge (\bar{x} = \bar{y})$$

where $\wedge$ is conjunction or "logical and".

For less than or equal we compare both the bounds for less than or equal like so,

$$[\underline{x}, \bar{x}] \leq [\underline{y}, \bar{y}] \text{ iff } (\underline{x} \leq \underline{y}) \wedge (\bar{x} \leq \bar{y})$$

Why equal and less than or equal? These relations are enough to define the others as needed and are called out explicitly in [7] which works towards defining a standard for interval arithmetic.

The last operation we are including in our interval library is the concept of the distance between two intervals. This is a scalar measuring the difference between the bounds of two intervals returning the largest difference. Symbolically this is,

$$\text{dist}(x, y) = \max\{|\underline{x} - \underline{y}|, |\bar{x} - \bar{y}|\}$$

which measures the degree to which two intervals "overlap". The larger this value, the more dissimilar the intervals are.

In C code these operations are defined to be,

```
1  unsigned char int_eq(interval_t x, interval_t y) {
2       return (x.a==y.a) && (x.b==y.b);
3  }
4
5  unsigned char int_le(interval_t x, interval_t y) {
6       return (x.a <= y.a) && (x.b <= y.b);
7  }
8
9  double int_dist(interval_t x, interval_t y) {
10      double a,b;
11      a = fabs(x.a - y.a);
12      b = fabs(x.b - y.b);
13      return (a>b) ? a : b;
14 }
```

where line 2 simply compares the bounds for equality and line 6 compares them for less than or equal. For int_dist() we first compute $|\underline{x} - \underline{y}|$ (line 11) and $|\bar{x} - \bar{y}|$ (line 12) and then return the larger of the two in line 13.

As a quick example of these functions consider two intervals,

$$x = [\texttt{-2.7182818284590451, 2.8182818284590452}]$$
$$y = [\ \texttt{2.518281828459 0449, 2.918281828459 0453}]$$

which overlap. With these, then, the function calls to our new operations return,

```
int_neg(x,y) = [-2.8182818284590452, 2.7182818284590451]
int_abs(x)   = [ 0.0000000000000000, 2.8182818284590452]
int_eq(x,y)  = 0 (false)
int_le(x,y)  = 1 (true)
int_dist(x,y) = 5.2365636569180900
```

The Python Version. Before we move on with our investigation of interval arithmetic let's quickly implement an interval class in Python. For this class we will only implement the basic arithmetic operations and leave extending the class to the exercises. The key to interval arithmetic is the ability to control the floating-point rounding mode. This is easily accomplished in C by including the fenv.h header file and compiling with the -frounding-math option. However, the Python interpreter is already compiled and does not offer direct access to the constants and functions defined in fenv.h. Fortunately, all is not lost, we are still able to access the underlying C library that contains these functions and use them to set the rounding mode as necessary. For this trick we thank Rafael Barreto for his blog post [8] which describes how to do this to set the rounding mode.

Jumping in, then, with our Python implementation we start with the beginning of the class,

```
 1  from math import exp
 2  from ctypes import cdll
 3  from ctypes.util import find_library
 4
 5  class Interval:
 6      libc = cdll.LoadLibrary(find_library("m"))
 7      FE_TOWARDZERO = 0xc00
 8      FE_DOWNWARD = 0x400
 9      FE_UPWARD = 0x800
10      FE_TONEAREST = 0
11
12      def __init__(self, a,b=None):
13          if (b == None):
14              self.a = a
15              self.b = a
16          else:
17              self.a = a
18              self.b = b
```

where lines 2 and 3 import standard Python libraries which will give us access to the C library so we can call the rounding functions. Line 1 imports the exponential function which we will use to generate infinity. The class starts on line 5. Line 6 sets a class-level variable, libc, to the C library which on Linux systems can be found by searching for "m" (hence -lm when compiling). We also define the rounding modes as class-level variables. The specific values were determined by a short C program which displayed them after including `fenv.h`. The constructor for the class starts on line 12 and takes up to two arguments. If only one is given the argument is assumed to be a real number and an interval is created from this number. Doing this simplifies the class by removing the need to consider mixed interval and float operations. If both arguments are given they are assumed to be the lower (a) and upper (b) bounds of the interval being defined. These are stored in `self.a` and `self.b`, respectively.

In order to round properly we need methods which set the rounding mode. These are defined next,

```
1    def __RoundDown(self):
2        self.libc.fesetround(self.FE_DOWNWARD)
3
4    def __RoundUp(self):
5        self.libc.fesetround(self.FE_UPWARD)
6
7    def __RoundNearest(self):
8        self.libc.fesetround(self.FE_TONEAREST)
```

where `_RoundDown()`, `_RoundUp()`, and `_RoundNearest()` are wrappers on calls to the C `fesetround()` function. Note the use of the double underscore prefix as Python notation for "private". We will call these methods whenever we need to change the rounding mode.

Next comes addition and subtraction. Instead of methods with those names we use the flexibility built into Python to define methods that work directly with the $+$ and $-$ operators. This means we can define expressions with intervals as if they were any other number or variable. The code is,

```
1    def __add__(self, y):
2        self.__RoundDown()
3        a = self.a + y.a
4        self.__RoundUp()
5        b = self.b + y.b
6        self.__RoundNearest()
7        return Interval(a,b)
8
9    def __sub__(self, y):
10       self.__RoundDown()
11       a = self.a - y.b
12       self.__RoundUp()
13       b = self.b - y.a
14       self.__RoundNearest()
15       return Interval(a,b)
```

where these methods, `_add_()` and `_sub_()`, will be called when objects of this class are encountered in arithmetic expressions involving $+$ and $-$. Since we are not implementing mixed operations both operands of the $+$ or $-$ must be of this class. Just as we did for the C library we set the rounding modes to down (lower bound) or up (upper bound) before calculating the new value. We return the new value as a new `Interval` object.

Multiplication is next (__mul__). This is directly analogous to the C version where we consider each case for the new lower and upper limit with proper rounding. The result is returned as a new interval object,

```
1    def __mul__(self, y):
2        self.__RoundDown()
3        a = min([self.a*y.a,self.a*y.b,self.b*y.a,self.b*y.b])
4        self.__RoundUp()
5        b = max([self.a*y.a,self.a*y.b,self.b*y.a,self.b*y.b])
6        self.__RoundNearest()
7        return Interval(a,b)
```

with line 2 setting rounding towards $-\infty$ before calculating all the products between the two intervals. It then selects the smallest of these and sets the lower bound to it. Line 4 changes to rounding towards $+\infty$ and repeats the calculation this time selecting the maximum value. We must do the multiplications twice because we changed the rounding mode. Line 6 returns to default rounding and the new object is returned in line 7.

Division, with reciprocal, and powers are all that remain to be implemented. Division is multiplication by the reciprocal, as before,

```
1    def recip(self):
2        if (self.a < 0.0) and (self.b > 0.0):
3            return Interval(-exp(1000.0), exp(1000.0))
4        self.__RoundDown()
5        a = 1.0 / self.b
6        self.__RoundUp()
7        b = 1.0 / self.a
8        self.__RoundNearest()
9        return Interval(a,b)
10
11   def __div__(self, y):
12       return self.__mul__(y.recip())
```

where division is called in line 12. The reciprocal method, not associated with an operator but not private, is used to return the reciprocal of the interval. If the interval contains zero (line 2) we return $[-\infty, +\infty]$ (line 3). Otherwise, we round down and calculate the lower bound, round up, calculate the upper bound, and finally return the new reciprocal object.

The last method we will implement is raising an interval to a positive integer power. This will mimic the C code as,

```
1    def __pow__(self, n):
2        if (n%2) == 1:
3            self.__RoundDown()
4            a = self.a**n
5            self.__RoundUp()
6            b = self.b**n
7        elif (self.a >= 0):
```

```
 8|              self.__RoundDown()
 9|              a = self.a**n
10|              self.__RoundUp()
11|              b = self.b**n
12|          elif (self.b < 0):
13|              self.__RoundDown()
14|              a = self.b**n
15|              self.__RoundUp()
16|              b = self.a**n
17|          else:
18|              a = 0.0
19|              self.__RoundUp()
20|              b = self.a**n
21|              t = self.b**n
22|              if (t > b):
23|                  b = t
24|          self.__RoundNearest()
25|          return Interval(a,b)
```

with each condition checked as before. Note that ** is exponentiation in Python.

Let's look at some examples using the class. We give a running interactive Python session with >>> the Python prompt followed by user input. Python output does not have the prompt and has been slightly intended for clarity,

```
>>> from Interval import *
>>> from math import sqrt
>>> x = Interval(-sqrt(2), sqrt(2)+0.1)
>>> x; x*x; x**2
    [-1.4142135623730951,  1.5142135623730952]
    [-2.1414213562373101,  2.2928427124746200]
    [0.0000000000000000,  2.2928427124746200]
>>> x*x*x; x**3
    [-3.2425692603699230,  3.4718535316173851]
    [-2.8284271247461903,  3.4718535316173846]
>>> x = Interval(sqrt(2), sqrt(2)+0.01)
>>> y = x**2
>>> x/y
    [0.6972118559681739,  0.7121067811865476]
>>> y/x
    [1.4042837765619229,  1.4342842730512142]
```

We first import the library and the square root function. We then define an interval, x, and show it along with $x \times x$ and x^2. Notice that the former gives too large an interval in this case because the interval contains zero. The more correct interval is the one returned by x^2. The same comparison is made between multiplying x by itself three times and calling the exponential. The multiplication leads to too large an interval. Re-defining x to be positive lets us demonstrate division thereby illustrating all the basic operations of the class.

In this section we explored and implemented basic interval arithmetic in C and Python. Next we consider functions of intervals. This is where we will run into the biggest reason, apart from performance issues due to repeated calculations, why interval arithmetic is not more widely used.

Properties of Intervals. We summarize this section by discussing some of the mathematical properties of intervals. The first few are familiar and mimic those of normal arithmetic. Things become more interesting when we get to the distributive property.

For any three intervals, $x = [\underline{x}, \bar{x}]$, $y = [\underline{y}, \bar{y}]$, and $z = [\underline{z}, \bar{z}]$, we have the following as true,

$$
\begin{aligned}
(x+y)+z &= x+(y+z) \quad \textit{(associative)} \\
(x-y)-z &= x-(y-z) \\
x+y &= y+x \qquad \textit{(commutative)} \\
x \times y &= y \times x \\
x+[0,0] &= x \qquad\qquad \textit{(additive identity)} \\
x \times [1,1] &= x \qquad\qquad \textit{(multiplicative identity)}
\end{aligned}
$$

which comes as no surprise. However, the distributive property causes problems. In general, for intervals,

$$x \times (y+z) \neq xy + xz$$

For example if $x = [1,2]$, $y = [1,2]$, and $z = [-2,-1]$,

$x \times (y+z)$	$xy + xz$
$[1,2] \times ([1,2]+[-2,-1])$	$[1,2] \times [1,2] + [1,2] \times [-2,-1]$
$[1,2] \times [-1,1]$	$[1,4] + [-4,-1]$
$[-2,2]$	$[-3,3]$

which is at first glance unexpected. This property is called *subdistributivity* and implies that for intervals we have,

$$x \times (y+z) \subseteq x \times y + x \times z$$

which is seen clearly in the example above because $[-2,2] \subseteq [-3,3]$ as it is a tighter bound. This means that in practice one needs to be careful with the form of an expression.

7.3 Functions and Intervals

Now that we know how to implement intervals we are ready to use them and be sure of accurate and tight bounds on our calculations. Or not. We must take a look at general functions of intervals and when we do we will see that a large elephant is indeed in the living room.

If a function of one variable is monotonic over a range this means that the function, $f(x)$, is either increasing or decreasing over the range. If this is the case, then interval arithmetic will work as expected because for the range of the interval the function will maintain (or reverse) the ordering of the bounds. For example, consider the square root function for positive intervals. Since the square root is monotonically increasing over any positive interval the output of $\sqrt{x}$ for an interval x will also be increasing as shown in Fig. 7.1.

Because of this consistent increase the tightest output interval will be simply $[\sqrt{\underline{x}},\sqrt{\bar{x}}]$. If the function is monotonically decreasing over the interval the result will be similar but with the bounds flipped to maintain a proper ordering. So, in general, for monotonically increasing functions of one variable ($f(x)$) we have,

$$f(x) = f([\underline{x},\bar{x}]) = [f(\underline{x}),f(\bar{x})]$$

while for monotonically decreasing functions ($g(x)$) we have,

$$g(x) = g([\underline{x},\bar{x}]) = [g(\bar{x}),g(\underline{x})]$$

where it should be noted that many elementary functions are of this form. This includes the exponential function, roots (eg. cube root), and logarithms. For exponentials and logarithms we can define the output explicitly,

$$a^{[\underline{x},\bar{x}]} = [a^{\underline{x}},a^{\bar{x}}], \ a > 1$$
$$\log_a[\underline{x},\bar{x}] = [\log_a\underline{x},\log_a\bar{x}]$$

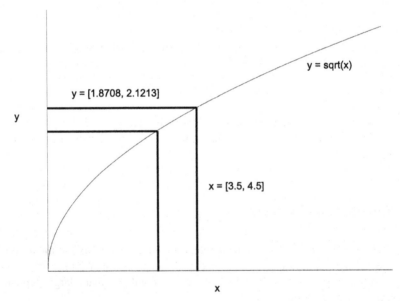

Fig. 7.1 An example of a monotonic function. Since the function $y = \sqrt{x}$ is always increasing for a positive input the bounds on any output interval are simply the function values at the lower and upper bounds of the input interval as shown for the interval $x = [3.5,4.5]$. This leads to the proper output interval of approximately $[1.8708,2.1213]$. If the function were monotonically decreasing over this interval the output would be the same but with the upper and lower bounds swapped so that the interval has the right ordering

What about functions that are not monotonically increasing or decreasing? In this case things become a bit more difficult. For example, what about the sine function? We assume the reader is familiar with the shape of the sine function. Clearly, it is sometimes increasing and sometimes decreasing. How to work with this and by extension the cosine?

One possibility is to use a Taylor series expansion. This is the solution presented in Daumas et al. [9] in terms of bounded functions using a parameter n to control the bounds where in the limit $n \to \infty$ the bounds tighten to the actual value of the function. Specifically, for the sine function [9] defines the output interval to be,

$$[\sin(x)]_n = [\underline{\sin}(\underline{x}, n), \overline{\sin}(\overline{x}, n)], \ x \subseteq [\frac{-\pi}{2}, \frac{\pi}{2}]$$

for a specified value of n. We'll see shortly what a good value of n might be. The functions $\underline{\sin}(\underline{x}, n)$ and $\overline{\sin}(\overline{x}, n)$ are themselves defined in terms of Taylor series parametrized by n and operating on scalar values (the bounds of x). The series expansions are,

$$\underline{\sin}(x, n) = \sum_{i=1}^{m} (-1)^{i-1} \frac{x^{2i-1}}{(2i-1)!}$$

$$\overline{\sin}(x, n) = \sum_{i=1}^{m+1} (-1)^{i-1} \frac{x^{2i-1}}{(2i-1)!}$$

where $m = 2n + 1, x \geq 0$, otherwise, $m = 2n$. In [9] the value of π itself is defined as an interval but for our purposes we will treat it as an exact quantity and note that the argument to our interval sine be restricted to the range given above.

Looking at the series expansions and the dependency upon the summation limit as a function of n we see where the cleverness lies in the definition. The bounds of the sine come from the number of terms used in the summation with the lower bound set by $m(n)$ and the upper set by $m(n) + 1$. Also, as $n \to \infty$ it is clear that the series will converge more and more to the sine of x.

Let's add this function to our interval library. The implementation is straightforward requiring a main function,

```
interval_t int_sin(interval_t x) {
    int n=8;
    interval_t ans;

    fesetround(FE_DOWNWARD);
    ans.a = int_sin_lo(x.a, n);
    fesetround(FE_UPWARD);
    ans.b = int_sin_hi(x.b, n);
    fesetround(FE_TONEAREST);
    return ans;
}
```

which follows our standard format of setting the rounding mode, first towards $-\infty$ for the lower bound and then towards $+\infty$ for the upper bound. Lines 6 and 8 set the output value by calling helper functions to compute the series value for the lower and upper bound. Note that line 2 sets n to eight. This value can be adjusted but eight gives good results. The helper functions are,

```
1  double fact(int n) {
2      return (n==1) ? 1.0 : n*fact(n-1);
3  }
4
5  double int_sin_lo(double x, int n) {
6      int i, m = (x < 0) ? 2*n : 2*n+1;
7      double ans=0.0;
8
9      for(i=1; i <= m; i++) {
10         ans += pow(-1,i-1) * pow(x,2*i-1) / fact(2*i-1);
11     }
12     return ans;
13 }
14
15 double int_sin_hi(double x, int n) {
16     int i, m = (x < 0) ? 2*n : 2*n+1;
17     double ans=0.0;
18
19     for(i=1; i <= m+1; i++) {
20         ans += pow(-1,i-1) * pow(x,2*i-1) / fact(2*i-1);
21     }
22     return ans;
23 }
```

where int_sin_lo() and int_sin_hi() call fact() to compute the factorial. We use the recursive algorithm for simplicity noting that we will only need less than a dozen terms in the series. The functions themselves are a literal translation of the formulas with m set appropriately when declared (lines 6 and 16). Note also that for int_sin_hi() we add one to m in the loop limit. Why do we not combine these two functions? If int_sin_hi() needs to calculate only one more term in the series expansion than int_sin_lo() why calculate everything twice? The answer lies in the guarantee implicit in intervals implemented in floating-point, namely, that the bounds on the interval will be certain to include the correct answer. When int_sin_lo() is called from int_sin() the floating-point rounding mode has been set towards $-\infty$. Therefore, all the calculations of the series terms will round in that direction thereby assuring us that rounding error will not shrink the interval we expect. Likewise, when int_sin_hi() is called the rounding mode it towards $+\infty$ for the same reason. This is why the calculation must be done twice.

Let's look at the output of our sine function and compare it to the "gold standard" present in the C math library. For,

$$x = [0.785\underline{2}981625, 0.785\underline{4}981625]$$

where the difference between the bounds is underlined, we get,

$$\sin(x) = [0.7070360663383\underline{567}, 0.7071774876944\underline{870}] \quad (\textit{interval sine})$$

while the sin() function from math.h produces the interval,

$$\sin(x) = [0.70703606633835\underline{76}, 0.70717748769448\underline{62}] \quad (\textit{math.h sine})$$

Looking at the underlined digits of each interval we see that our sine function does indeed produce an interval which will be sure to encompass the "true" value.

The formulas for the interval version of cosine are,

$$[\cos(x)]_n = [\underline{\cos}(\bar{x}, n), \overline{\cos}(\underline{x}, n)], \ x \subseteq [0, \pi]$$

and,

$$\underline{\cos}(x, n) = 1 + \sum_{i=1}^{m+1} (-1)^i \frac{x^{2i}}{(2i)!}$$

$$\overline{\cos}(x, n) = 1 + \sum_{i=1}^{m} (-1)^i \frac{x^{2i}}{(2i)!}$$

again where $m = 2n + 1, x \geq 0$, otherwise, $m = 2n$. We leave the implementation, in C and Python, to the exercises.

So far in this section we have examined interval versions of elementary functions. What about general functions of x? Let's take a look and we will soon see that this is the elephant mentioned above. Applying interval arithmetic to general functions leads to the *dependency problem* (or *dependency effect*) which is the bane of intervals.

This problem shows itself when applying intervals to general functions where the interval variable appears more than once in the expression. The most commonly cited example of this is the evaluation of the expression $x - x^2$ for $x = [0, 1]$. Another way to write this expression is $x(1 - x)$. Let's look at how these two compare when we evaluate them,

$x(1-x)$	$x-x^2$
$[0, 1] \times (1 - [0, 1])$	$[0, 1] - [0, 1]^2$
$[0, 1] \times [0, 1]$	$[0, 1] - [0, 1]$
$[0, 1]$	$[-1, 1]$

where we clearly have differing answers. We also have that the interval for $x(1 - x)$ $\subseteq x - x^2$ which is exactly what we would expect due to the subdistributivity of intervals. However, while neither answer is technically incorrect, since the proper answer lies inside of each interval, neither answer is satisfactory because of the dependency problem.

The expression we are evaluating is a downward facing parabola with a maximum value of 0.25 at $x = \frac{1}{2}$ and minimum values of zero at $x = 0$ and $x = 1$. Therefore, over the given range, $x = [0, 1]$, we would expect the output interval to be $[0, \frac{1}{4}]$. While this interval is within the range of both of the intervals above we do not find the narrow interval unless we remove the dependency problem by rewriting the expression so that x appears only once. If we complete the square of $x - x^2$ we can rewrite the original expression as $\frac{1}{4} - (\frac{1}{2} - x)^2$ in which x appears only once. If we evaluate this expression over $x = [0, 1]$ we get,

$$\frac{\frac{1}{4} - (\frac{1}{2} - x)^2}{0.25 - (0.5 - [0,1])^2}$$
$$0.25 - [0,0.25]$$
$$[0,0.25]$$

which is the narrowest interval in this case.

The dependency problem is likely the primary reason interval arithmetic is not more widely used. It is often impossible to rewrite an expression so that the variable appears only once. When this is the case it is possible to introduce auxiliary intervals whose union results in the original interval and then apply the expression to each of these intervals. This *interval splitting* will result in less of a dependency problem. Let's look at an example of this from [9] where the expression under consideration is $2x - x$ with $x = [0,1]$. If we simply evaluate the expression we get $[-1,2]$ as the answer when the narrowest interval is naturally $[0,1]$. The observation key to interval splitting is that while $x = [0,1]$ does not lead to the narrowest interval we can split the interval into two parts, $x_1 = [0,0.5]$ and $x_2 = [0.5,1]$, and evaluate each of these in turn. In that case we get $[-0.5,1]$ for x_1 and $[0,1.5]$ for x_2. The union of these two intervals is $[-0.5,1.5]$ which is closer to the expected answer of $[0,1]$. If the intervals are divided further and further the resulting answer will approach $[0,1]$ which we can see if we evaluate $x_1 = [0,0.05]$ to $x_{20} = [0.95,1]$ which gives $[-0.05,0.1]$ up to $[0.9,1.05]$ the union of which is $[-0.05,1.05]$. While interval splitting can help deal with the dependency problem it introduces a significant amount of overhead to the calculations. This is why our simple interval C library and Python class do not implement interval splitting even though more complete libraries do.

7.4 Implementations

In this section we look at two implementations of interval arithmetic, one for C and the other for Python. For C perhaps the most widely used library is the Multiple Precision Floating-point Interval library or MPFI for short [10]. This library is built on top of the widely used GNU GMP and MPFR libraries for multiple precision floating-point numbers. This means that in addition to interval arithmetic the MPFI library supports more than the standard IEEE 754 32-bit and 64-bit floating-point. For our purposes we will ignore this important and useful feature and instead focus on the interval support MPFI provides. Using MPFI requires us to install GMP (*http://gmplib.org/*) and MPFR (*http://www.mpfr.org/*). Once these are installed with the typical "`configure; make; make install`" sequence of Unix instructions we can download MPFI (*http://perso.ens-lyon.fr/nathalie.revol/ software.html*). Building MPFI follows the same sequence of steps as GMP and MPFR.

Let's start with a simple example that defines two intervals and then prints them on standard out. This will teach us how to define and initialize intervals and how to print them. Please note that there is an extensive API for this library and that we are only using a small part of it. Our initial program, then, is,

```
 1 #include <stdio.h>
 2 #include <mpfr.h>
 3 #include <mpfi.h>
 4 #include <mpfi_io.h>
 5
 6 int main() {
 7     mpfi_t x, y;
 8
 9     mpfr_set_default_prec(64);
10
11     mpfi_init(x);
12     mpfi_set_str(x, "[-1.0, 1.0]", 10);
13
14     mpfi_init_set_str(y, "[0,2]", 10);
15
16     printf("x = ");
17     mpfi_out_str(stdout, 10, 0, x);
18     printf("\n");
19
20     printf("y = ");
21     mpfi_out_str(stdout, 10, 0, y);
22     printf("\n");
23
24     mpfi_clear(x);
25     mpfi_clear(y);
26 }
```

where lines 1 through 4 include necessary header files. Note that the documentation for MPFI recommends including stdio.h before the remaining header files. Line 7 declares two interval variables, x and y. This declares the variables but does not initialize them. They must be initialized before they are used. Line 9 sets the default precision to 64 bits to mimic a double. Note that this is a call to the MPFR library. Lines 11 and 12 show one way to initialize an interval variable and set its value from a string. We will use strings to set values though there are many options in the API. For mpfi_set_str() the first argument is the variable, the second is a string which defines the interval in the way we have been displaying them throughout this chapter, and the last argument is the base in which the interval is defined (2 through 36). Line 14 combines lines 11 and 12 into a single initialize and set function. The arguments are the same as line 12. We then display the intervals using mpfi_out_str() in lines 17 and 21. The first argument is the output file stream, the second is the base, the third is the number of digits (0 says use the precision of the variable) and the fourth argument is the interval variable itself. Lastly, lines 24 and 25 clear the memory used by x and y. While not necessary in this case because the program is about to exit it is necessary before re-assigning these variables.

This program should be compiled with,

```
gcc mpfi_ex.c -o mpfi_ex -lmpfi -lmpfr
```

assuming mpfi_ex.c to be the file name. Note that MPFI installs itself in /usr/local/lib which might require adding that path to the LD_LIBRARY_PATH environment variable.

Now that we know how to initialize, set, and display intervals in MPFI let's look at basic arithmetic functions. These are,

```
int mpfi_add(mpfi_t ROP, mpfi_t OP1, mpfi_t OP2)
int mpfi_sub(mpfi_t ROP, mpfi_t OP1, mpfi_t OP2)
int mpfi_mul(mpfi_t ROP, mpfi_t OP1, mpfi_t OP2)
int mpfi_div(mpfi_t ROP, mpfi_t OP1, mpfi_t OP2)
```

where ROP is the variable to receive the result and OP1 and OP2 are the operands. Note that division using an interval containing zero will result in $[-\infty, +\infty]$. The return value indicates whether or not endpoints of the interval are exact.

MPFI is powerful but a bit cumbersome to use because of its C language heritage. There is a C++ version as well which we are not exploring here. Let's look at our dependency problem example again and implement it using MPFI. This example computes $x(1-x)$ and $x - x^2$ for $x = [0,1]$. The C code is,

```
 1  int main() {
 2      mpfi_t x, one, t, y;
 3
 4      mpfr_set_default_prec(64);
 5      mpfi_init_set_str(x, "[0, 1]", 10);
 6      mpfi_init_set_str(one, "[1,1]", 10);
 7
 8      mpfi_init(t);
 9      mpfi_sub(t, one, x);
10      mpfi_init(y);
11      mpfi_mul(y, t, x);
12
13      printf("y = ");
14      mpfi_out_str(stdout, 10, 0, y);
15      printf("\n");
16
17      mpfi_clear(y);
18      mpfi_clear(t);
19      mpfi_clear(one);
20
21      mpfi_init(t);
22      mpfi_mul(t, x, x);
23      mpfi_init(y);
24      mpfi_sub(y, x, t);
25
26      printf("y = ");
27      mpfi_out_str(stdout, 10, 0, y);
28      printf("\n");
29  }
```

where we see that it is necessary to preserve the intermediate temporary variables so that they can be released properly without leaking memory. Specifically, line 5 initializes x while line 6 initializes the constant interval, one. The expression $x(1-x)$ is calculated in lines 8 through 11. First, $t = 1 - x$ is calculated in line 9 and then the final answer is found as $y = t \times x$ in line 11. This is the first y output (line 14). Lines 17 through 19 release memory so that we can reuse y and t to calculate $x - x^2$. This expression is calculated in lines 21 through 24 and output in line 27. The output,

```
y = [0,1.0000000000000000000]
y = [-1.0000000000000000000,1.0000000000000000000]
```

is precisely what we would expect it to be for $x(1-x)$ and $x-x^2$.

The functionality of MPFI is available in Python via the mpmath [11] library. This library is built on the same foundations as MPFI and is quite powerful. As with MPFI, we will only look at the simplest of uses with intervals. To install mpmath on Ubuntu use the command,

```
sudo apt-get -y install python-mpmath
```

with suitable repository access for other Linux distributions or build it from the website.

Once the library is installed we can access it easily from the Python command prompt. The following console session shows some of mpmath's basic interval abilities,

```
 1 │ >>> from mpmath import mpi
 2 │ >>> x = mpi('0','1')
 3 │ >>> print x
 4 │     [0.0, 1.0]
 5 │ >>> print x*(1-x)
 6 │     [0.0, 1.0]
 7 │ >>> print x-x**2
 8 │     [-1.0, 1.0]
 9 │ >>> y = mpi('-1','1')
10 │ >>> print y**2
11 │     [0.0, 1.0]
12 │ >>> print y*y
13 │     [-1.0, 1.0]
14 │ >>> print 1/y
15 │     [-inf, +inf]
16 │ >>> print 1/x
17 │     [1.0, +inf]
```

where the Python prompt is >>> and Python output is indented for clarity.

In line 1 we import the interval module (mpi) from the larger mpmath library. Line 2 defines a new interval and stores it in x. Note that the bounds are given as two separate arguments and may be given as strings or as numbers. If given as numbers they are interpreted as Python floats (64-bits) while strings are interpreted as exact decimals. Line 5 prints $x(1-x)$ with the expected output while line 7 prints $x-x^2$, again with the expected output. Line 9 defines y to be an interval containing zero. Lines 10 through 15 display the results of several simple calculations on y. Note that y^2 (line 10) returns an interval different than $y \times y$ (line 12) but positive, as in our C library above. Note that line 14 returns $[-\infty, +\infty]$ as the reciprocal of y since it contains zero while line 16 prints the reciprocal of x which has zero as a lower bound. In this case the resulting interval is $[0, +\infty]$.

The bounds of an interval are acquired by the a and b properties on the object. Additionally, there are also mid and delta properties for the midpoint of an interval and for the difference between the endpoints,

```
1 >>> from mpmath import mpi
2 >>> x = mpi('0.41465', '0.5')
2 >>> x.a
2     mpi('0.41464999999999996', '0.41464999999999996')
2 >>> x.b
2     mpi('0.5', '0.5')
2 >>> x.mid
2     mpi('0.45732499999999998', '0.45732499999999998')
2 >>> x.delta
2     mpi('0.085350000000000037', '0.085350000000000037')
```

with x.a returning the lower bound, x.b returning the upper, x.mid returning the middle position between the lower and upper bounds, and x.delta to calculating the range of the interval. Note, the return values are themselves intervals even if the lower and upper bounds are the same.

This section took a quick look at two common interval libraries to extend C and Python. Intervals have been added to many other languages and sometimes the compiler used will support intervals. For example, Fortran has had interval packages for some time. More recent languages like R and Matlab also support intervals.

7.5 Thoughts on Interval Arithmetic

Interval arithmetic is most useful when uncertainty is part of the data. With intervals, and proper rounding, we have assurances that the correct answer lies within a particular range. For basic operations intervals work well even if they may involve significantly more low-level calculations than floating-point numbers on their own. For that reason, unless truly necessary, intervals are generally not recommended in situations where high-performance is required. Packages like MPFI and mpmath use powerful underlying libraries with arbitrary precision floating-point and this can add much to creating intervals that are highly accurate and as small as is possible.

Floating-point arithmetic is standardized via the IEEE 754 standard. IEEE 754 defines the number of bits for each precision, including single (32-bit) and double (64-bit), as well as rounding modes, comparisons, etc. The IEEE organization, as of this writing (Fall 2014), is developing IEEE P1788 (Standard for Interval Arithmetic) as a project [12]. This standard will address the fact that currently intervals are implemented in many different ways and without a consistent set of algorithms. While not yet complete and adopted, the standard is worth remembering as it may go a long way towards increasing the use of interval arithmetic. A good summary of interval arithmetic, along with a discussion of its uses in scientific and mathematical research, can be found in [13].

Because of the difficulty involved in properly applying interval arithmetic to general functions and expressions one must be careful and thoughtful when using intervals to avoid falling into a situation where the uncertainty grows too large. Recall that unlike standard floating-point computation, the interval $[-\infty, +\infty]$ is a valid return value for an interval expression, albeit, not a particularly useful one. However, if uncertainty and bounds on it are crucial to the task at hand, then interval arithmetic may be the ideal choice.

7.6 Chapter Summary

In this chapter we took a practical look at interval arithmetic. We defined what we mean by intervals and interval arithmetic, summarized the history of intervals, and explained basic operations on intervals. We implemented an interval library in C and Python which included some elementary functions. We then discussed the dependency problem, with examples, and mentioned interval splitting as a technique for minimizing the dependency problem. Lastly, we looked at some implementations of interval arithmetic and offered some thoughts on interval arithmetic and its utility.

Exercises

7.1. Add `int_lt()`, `int_gt()`, `int_ge()`, and `int_ne()` functions to test for $<$, $>$, $\geq$, and $\neq$ using `int_eq()` and `int_le()`. **

7.2. Add exponential (e^x), square root ($\sqrt{x}$) and logarithm functions (any base) to the C interval library. **

7.3. The Python interval class only contains basic operations. Complete the class by adding the missing operations found in the C library. **

7.4. Extend the Python interval library to support mixed scalar and interval arithmetic for $+$, $-$, $\times$, and $/$. (Hint: look at how methods like `__radd__()` might be used) **

7.5. Extend the Python `__pow__()` method to proper handle negative integer exponents.

7.6. Extend the C library by adding cosine using the formulas in Sect. 7.3. *

7.7. Extend the Python class by adding sine and cosine using the formulas in Sect. 7.3. *

7.8. Add interval splitting to the Python interval class in order to minimize the dependency problem. Automatically split intervals into n parts and evaluate a function on each of those parts reporting the union of them when complete. ***

References

1. Bevington, PR., Robinson, DK. Data reduction and error analysis for the physical sciences. Vol. 336. McGraw-Hill (1969).
2. Young, RC., The algebra of multi-valued quantities, Mathematische Annalen, 1931, Vol. 104, pp. 260–290.
3. Dwyer, PS., Linear Computations, J. Wiley, N.Y., 1951.

4. Sunaga, T., Theory of interval algebra and its application to numerical analysis, In: Research Association of Applied Geometry (RAAG) Memoirs, Geujutsu Bunken Fukuy kai, Tokyo, Japan, 1958, Vol. 2, pp. 29–46 (547–564); reprinted in Japan Journal on Industrial and Applied Mathematics, 2009, Vol. 26, No. 2–3, pp. 126–143.

5. Moore, RE. Automatic error analysis in digital computation. Technical Report Space Div. Report LMSD84821, Lockheed Missiles and Space Co., 1959.

6. Hickey, T., Ju, Q. and van Emden, MH., Interval Arithmetic: From principles to implementation. Journal of the ACM (JACM) 48.5 (2001): 1038–1068.

7. Bohlender, G., Kulisch, U., Definition of the Arithmetic Operations and Comparison Relations for an Interval Arithmetic Standard. Reliable Computing 15 (2011): 37.

8. Barreto, R., Controlling FPU rounding modes with Python, http://rafaelbarreto.wordpress.com/2009/03/30/controlling-fpu-rounding-modes-with-python/ (accessed 07-Nov-2014).

9. Daumas, M., Lester, D., Muoz, C. Verified real number calculations: A library for interval arithmetic. Computers, IEEE Transactions on 58.2 (2009): 226–237.

10. Revol, N., Rouillier, F., Multiple Precision Floating-point Interval Library. http://perso.ens-lyon.fr/nathalie.revol/software.html (retrieved 15-Nov-2014) (2002).

11. Johansson, F. et al. mpmath: a Python library for arbitrary-precision floating-point arithmetic (version 0.14), February 2010. http://code.google.com/p/mpmath (accessed 16-Nov-2014).

12. Kearfott, R. An overview of the upcoming IEEE P-1788 working group document: Standard for interval arithmetic. IFSA/NAFIPS. 2013.

13. Kearfott, R. Interval computations: Introduction, uses, and resources. Euromath Bulletin 2.1 (1996): 95–112.

Index

© Springer International Publishing Switzerland 2015

R.T. Kneusel, *Numbers and Computers*, DOI 10.1007/978-3-319-17260-6

Printed in the United States
By Bookmasters